Agricultural Geography

AGRICULTURAL GEOGRAPHY

Girish Chopra

Commonwealth

Agricultural Geography

Edition : 2024

ISBN 81-7169-997-9

Published by :
COMMONWEALTH PUBLISHERS PVT. LTD.,
4831/24, Prahlad Street,
Ansari Road, Darya Ganj,
New Delhi - 110002
Ph. : 23275511, 23257835
e-mail : commonwealthbooks@gmail.com

Typesetting by :
Arnav Computers
Delhi

Printed at :
Roshan Offset Printers
Delhi

Preface

Agriculture uses about a third of the land on the planet and occupies the lives of about two and a half billion people. Around ten to twelve thousand years ago, human began to domesticate plants and animals for food. Before this first agricultural revolution, people relied on hunting and gathering to obtain food supplies. While there are still groups of hunters and gatherers in the world, most societies have switched to agriculture.

The beginnings of agriculture did not just occur in one place but appeared almost simultaneously around the world, possibly through trail and error with different plants and animals or by long term experimentation. Between the first agricultural revolution thousands of years ago and the 17th century, agriculture remained pretty much the same. In the seventeenth century, a second agricultural revolution took place which increased efficiency of production as well as distribution which allowed more people to move to the cities as the industrial revolution got under way. Farming in the twentieth century has become highly technological in more developed nations with geographical technologies like GIS, GPS, and remote sensing while less developed nations continue with practices which are similar to those developed after the first agricultural revolution, thousands of years ago.

Today, about 45 per cent of the world's population makes their living through agriculture. The proportion of the population involved in agriculture ranges from about 2 per cent in the United States to about 80 per cent in some parts of Asia and Africa. There are two types of agriculture, subsistence and commercial. The subsistence farmers produce only enough crops to feed their families. While the primary purpose of commercial farmers is to sell their product at market.

The purpose of this book is to provide the reader with an in-depth knowledge of agricultural geography. It supplies an introduction to the major aspects of agricultural geography: the origin and spread of agriculture; issues of land and natural resources management; agricultural and sustainable development; agricultural research and planning and the prospects of agricultural baseline. The content of the book will be immensely useful to students, teachers, researchers and all who interested in the subject.

Girish Chopra

Contents

1

Agriculture and Sustainable Development

Agriculture is the dominant use of land in all countries. Use of land for this purpose can affect the environmental sustainability of the natural resource base which in turn has implications for the economic viability of farming. Together, these considerations raise important questions about the sustainability of agriculture and the implementation of policy prescriptions, including budgetary reforms, for both environmental protection and economic development.

The International Institute for Sustainable Development (IISD), through its Board of Directors and its mandate to be a catalyst for change, launched a study on sustainability of agriculture. The intent was to establish a framework for analysis of current sustainable development issues that would lead to recommendations for change. The study was complementary to that in the United States, dealing with local issues on the Canadian prairies. The analytical framework developed can be applied throughout the entire Great Plains Region and other semi-arid regions of the world.

The study selected the Great Plains—a vast region extending from the Canadian prairies through the Western

United States—as its study area. Interest in the sustainability of agriculture and the environment throughout this region is critical for the realization of economic, social and environmental benefits. In the United States, for example, the Western Governors' Association is preparing information and strategies to foster sustainable development within its region. This initiative will require addressing conservation challenges, building partnerships among stakeholders and developing guidelines for sustainability. While this initiative has attracted attention in Canada and led to cooperation with the Association, there is a continuing need to deal with sustainable development issues on the Canadian prairies.

The sustainability of agriculture on the prairies should be prefaced by an outline of the historical development process. Sustainability in a semi-arid area characterized by highly variable rainfall has been greatly affected not only by the original method of land distribution but also by the development of technology. It is also affected by those fluctuations in net returns that are typical of an agriculture dependent upon the world market for a major share of the output produced. The initial development of agriculture on the prairies was largely associated with the Hudson's Bay Company. In a charter granted in 1670, the company was given a monopoly on the fur trade in Rupert's Land, and was made responsible for creation of this trade using its own resources and for establishing a colony. Operators of the fur trading posts attempted to grow produce in order to provide a more varied diet than was available from local sources.

The colony was located at the forks of the Red and Assiniboine rivers. The difficulties faced by settlers in the colony were indicative of those which were to continue to be experienced by agriculturists on the prairies. These difficulties included frost, drought, floods and grasshoppers. The limited output of produce found a ready market at the time among the fur traders. It is said of the period when the country was under the Hudson's Bay Company control that agriculture developed as much as could be expected considering the climatic conditions, the difficulties of transportation and the concentration on the fur trade.

The prairies were a vast grassland with tree cover limited to river valleys and high elevations, due to the rampant prairie fires. Two expedition teams explored the area during the 1857-60 period. One was led by Capt. John Palliser and the other by S.J. Dawson. Both expeditions independently reached the same conclusion, determining that the interior of the west was composed of an area somewhat triangular in shape with the base running along the 49th parallel from about present day Killarney, Manitoba to Pincher Creek, Alberta with the apex being in the proximity of Saskatoon, Saskatchewan. This area they considered to be an extension of the Great American Desert, with a fertile belt of excellent soil lying to the north. On the other hand, J. Macoun,in 1882, stated that the so-called Palliser triangle was not as arid as previously believed but was "well suited to agriculture." The truth lay somewhere in between. The former view was based on years of low rainfall. The latter focussed on particularly fertile parts of the Souris Valley and the Regina plains during the high rainfall years of the late 1870s, and contained a substantial degree of self-induced optimism.

The National Policy adopted following Confederation encompassed several objectives. It included tariffs to protect eastern industry, a railway or railways to join Canada together and protect the west from being absorbed by the Americans, and settlement to provide traffic for the railways and a market for the products of eastern industry. Settlement was key and settlers were essential to the process. The lands policy adopted was that which prevailed in the Plains region of the United States. The free homestead system, preemptions, school lands and railway land grants were all based upon previous American experience. Railway and school lands later proved invaluable. They allowed the original "free" homestead to consistently expand as an economic unit under the dry land farming conditions that prevailed on the prairies.

Settlers were attracted to the prairies by the opportunity to own land and also by what was perceived as a chance to improve their economic position. Homesteading was said to be a gamble in which the federal government bet 160 acres with a settler against $10 that he could not stay on the homestead for six months for each of three years without starving to death. Over forty percent of the settlers left, effectively losing the bet. While such an adverse social outcome is deplorable, it must be recognized that most settlers arrived with few finances and had to complement their meager earnings from the land with income from railway construction or other local activity. Relatively few of the original settlers had dry land farming experience.

The settlers who persisted demonstrated their ability to adapt to the local situation. The sustainability of an agriculture based on grain was early placed in doubt. The 1880s were characterized by drought with crops sown on stubble providing little return. While the need to set aside seed and feed in "good" years was recognized it was discovered that allowing land to lie fallow enabled it to absorb enough moisture to provide at least a sustaining crop the following year. This idea was widely publicized by the Experimental Farms established by the federal government.

The existing crop varieties, particularly those of wheat, had a growing period which tended to exceed the number of frost free days. Research provided varieties more closely adapted to the local environment and Red Fife was replaced in 1911 by Marquis, the flagship variety for "quality" in Canadian wheat for the following seventy years.

The initial period covering from about 1880 to 1900 can be taken as the first stage in a sustainable development learning curve continuum which persists to this day. Oxen and the walking plow were followed by horses and the gang plow. The scythe and cradle were replaced by the horse drawn binder. The threshing machine was adopted at an early stage along with the steam engine. The latter required a crew of men to operate. The Crow's Nest Pass Agreement of 1897 ensured that rail rates on grain shipments out and agricultural inputs into the prairies would foster development. Meanwhile, the Manitoba Grain Act of 1900 protected the interests of grain producers when marketing their grain.

The initial objective was to develop an agricultural industry which would yield an income sufficient to support the local population and others at a distance while providing the necessities of life. Though the emphasis was on grain for export, mixed farming was practiced since producers learned early in the process the desirability of having livestock to provide sustenance during drought periods. They had the support in this initiative of local governments. The animal culture followed, and in combination with the scale of grain operations obviated any need for concern for sustainability other than to overcome the effects of the climate.

The period 1900 to 1910 was one of great expansion. In response to the lure of cheap land and the advertising by government and railways alike, a great influx of settlers took place on the prairies, totalling 251,594 homestead entries. The methods of farming continued as before while the area in farms increased rapidly. The expansion was facilitated by the extension of the railway network. Much effort was expended by the Department of the Interior, the provincial departments of agriculture and the Canadian Pacific Railway to upgrade the livestock herds and to foster mixed farming practices. The period was one of relative prosperity even though most of the expansion occurred in the semi-arid areas. However, in the expansion, little heed was paid to the quality of land being opened up for settlement, something sure to haunt the prairies in later years.

During the period 1910 to 1930, further expansion of settlement occurred. Essentially, by the end of this period the prairie lands had been occupied. Service industries became established. The war years of 1914-18 were marked by an increasing demand for wheat and attractive prices prevailed.

However, drought years continued to be interspersed with those of more liberal rainfall. Livestock herds continued to expand. Developments in farm machinery, particularly toward the end of the period with the advent of the farm tractor and improved threshing machines, enabled land to be worked with fewer people than was the case when the horse was needed for power. Less land was required for pasture and for the production of feed. Producers were in an expansive frame of mind and average farm size increased. However, during the period, it became evident that some areas being farmed in southern Alberta (outside the irrigation area) and in southwestern Saskatchewan were not suited to the production of grain and should be returned to grass.

The Royal Commission examining the situation in 1921 concluded that a soil survey should be undertaken. The lack of information on soil capability prior to settlement was now being made painfully evident by the distress of producers and the abandonment of farmsteads. The emphasis, nonetheless, continued to be on production with wheat as king.

Disaster struck the prairies in the 1930s. Much of the expansion of the earlier years was debt financed. The collapse in wheat prices brought economic ruin not only to individual producers but also to many of their organizations and suppliers. Indeed, the period 1930-1940 has been spoken of as the "dirty thirties" and "the winter years". The decade was marked by seven years of drought, ending in 1937 when the crop failure in Saskatchewan was almost complete. Accompanying the drought came grasshoppers and army worms which destroyed much of what little was produced. Waters hortages in the southern areas were common.

The detrimental effects of the plow, intensive farming practices made possible by mechanization, and the leaving of large tracts in summer fallow became evident as the high winds of the period gave rise to extensive soil drifting. So dark did the sky become at times from dust that the hens retired to roost in the middle of the day. There began an exodus of 250,000 people from the dry areas in the south to the more favourable areas in the north resulting in deserted farmsteads and broken dreams. The social cost was enormous. The combination of events was not foreseen and influenced those involved for a lifetime. Gray states that while the drought itself could be blamed on nature, the "Dust Bowl" was man made. Millions of acres had been broken by the plow which should never have been broken at all. The land was only suitable for grazing. The dry topsoil began to blow in 1931 and it blew summer and winter until 1938. Producers realized that change was required in their farming practices as the sustainability of agriculture in large areas of the southern prairies came into doubt.

The situation called for drastic action beyond the extension of relief to the stricken areas. Such action was rendered feasible by a reorientation of the work of those researchers who had previously devoted their attention to increasing production. The Dominion Rust Research Laboratory, established in 1925, produced new varieties of grain resistant to disease, such as rust. These new varieties were of great benefit to the prairies.

The Dominion Experimental Farms had developed new methods of farming. Strip farming provided some protection from the wind. The resourcefulness of producers was tested. They recognized the need for trash cover on their fields, which required new types of tillage equipment. The answer was the Noble blade developed in 1934 and other similar trash maintaining cultivators invented by producer ingenuity. Judicious use of the oneway disc also allowed maintenance of trash cover in certain areas.

The federal government's response to the situation had far reaching implications. The Prairie Farm Rehabilitation Act was passed and an Administration formed (PFRA) in 1935. Technology was applied to solve the problems, even to the distribution of bait for grasshopper control. Drought tolerant grasses and legumes were introduced. The PFRA in cooperation with the Experimental Farms provided 48 experimental substations to teach producers how to control soil drifting. Tree planting was promoted. Almost three million acres in western Saskatchewan and Alberta were restored to grass and community pastures were formed.

Dugouts were made to provide on-farm water supplies. Local irrigation schemes were promoted and advice provided on how to store water for livestock.

Farmers were relocated from the disaster areas to others with greater potential. The Canadian Wheat Board, established in 1935, provided a degree of stability to wheat prices. To offset crop disasters, the Prairie Farm Assistance Act was passed in 1939 as the forerunner to the crop insurance that became available 20 years later. These imaginative actions of the federal government had their counterparts in the United States, where similar disasters affected agriculture on the western plains during the period.

Regardless of the problems associated with the decade, improvements in farm tractors and combines led to greater mechanization on farms. The tractor proved to be a good replacement for "horse-power" under drought conditions. Combines, developed to be suitable for the average farm, enabled straw to be spread over the surface of the land. Mechanization enabled more to be done by less people, thereby adding to underemployment in agriculture. A new era in agriculture began in 1938 with the change in the weather cycle. The lessons of the drought years were not to be soon forgotten. Indeed, attention continued to be placed on maintenance of trash cover as the ingenuity of producers gave rise to the disker, which enabled large areas to be covered with a tillage implement that also had the capacity to be utilized for seeding. The self propelled combine came into its own while farm trucks provided the means for transporting grain to the elevator.

These developments allowed larger acreages to be farmed with fewer workers. Even so, farm labourers were in scarce supply during the war years as the young men on the prairies were drawn away to the armed services. In the services, their experience with machinery became invaluable to the air force, army and navy. The taunts of those who classed them as "stubble jumpers" assisted by "beans and old clothes" were quickly cast aside as a result of their performance.

While the early war years, i.e., 1939-42, were notable for the institution of delivery quotas on wheat, they were also marked by the encouragement of an expansion in livestock output, particularly hogs, in support of the war effort. The feed freight subsidy was introduced in 1941 as a "temporary" measure to increase livestock production outside the prairies. Following the termination of hostilities, the need to replenish the food supply in Europe resulted in all out grain production. This situation was short-lived as the war-devastated areas quickly recovered under the stimulus of the Marshall Plan. At the same time, the industrialization of Canada, which was stimulated by the war, persisted, resulting in agriculture's contribution to the economy continuing to fall proportionately.

The 1950s and 1960s were marked by recurrent surpluses in grain production. To reduce the storage costs experienced by grain producers, the Temporary Wheat Reserves Act was introduced in 1954 and was followed by the two-price wheat program in 1967. The "dripping" '50s gave way to the occasional droughts of the '60s. New markets for wheat opened up as China and the Soviet Union became major customers.

Meanwhile, farms became more and more mechanized and human energy was replaced by electric power and more versatile farm equipment. The surplus in grain production became acute by 1970 when operation LIFT, Lower Inventory for Tomorrow, was introduced. The restrictive quotas had earlier been based largely on cultivated acreage giving rise to extensive grain production. With LIFT, an emergency program to reduce output, delivery quotas for wheat were based on acres in summer fallow, a development not in the best interest of sustainable agriculture as some land was summer fallowed two years in succession.

The Task Force on Agriculture Report in 1969 drew attention to the problems inherent in agriculture on the prairies. It suggested that the emphasis on wheat should be partially diverted to the production of barley, with more consideration given to the production of forage for livestock. The conclusions of the Task Force were based on perceived market opportunities. However, shortages, real or perceived, in the world output of grains and oilseeds in 1973/74 gave rise to an increase in demand which set off, in combination with rising oil prices, a round of inflation the effects of which remain today in terms of land prices exceeding those which can be supported by net returns.

While a temporary stimulus had been given to hog production on the prairies by the inability to market grain, the increase was quickly dissipated as the returns from grain became more lucrative. At the same time passage of the Western Grain Stabilization Act in 1974 provided a measure of income stability to grain producers in general, thereby improving the economic stability of the prairie region. In addition, the crop insurance which had replaced the Prairie Farm Assistance Act offset much of the impact of yield variability on net returns. Meanwhile, the effects of the below-cost freight rates on grain for export became increasingly obvious, the railways allowing their grain dependent branch lines and grain movement capability to deteriorate. The federal government and the Canadian Wheat Board intervened to overcome this decline in capability, their actions leading to the passage of the Western Grain Transportation Act in 1983.

What could be considered to be a federal obsession with the prairie grain economy had detrimental effects on the livestock industry. These were exacerbated by the quota policy of the Canadian Wheat Board. Greater livestock production could have offset the tendency to develop a grain monoculture on the prairies. The reduction in risk brought about by the government policies favoured development of larger farms. The emphasis on grain was further accentuated by the 1980 forecast of the Canadian Wheat Board for an increasing world demand for Canadian grain. Meanwhile, technology also favoured a grain monoculture.

The multiplication of farm tractor horsepower and the increasing size and sophistication of implements enabled greater acreages to be operated with less labour. Herbicides, pesticides and fertilizers provided the means for increasing yields, the former allowing a reduction in the area devoted to summer fallow. While wheat remained king, oilseeds and special crops, particularly canola, provided a measure of diversity in production.

The family farm, which earlier had been considered a way of life and which federal legislation was designed to promote, had become a business requiring a high degree of management skill. It was highly mechanized, highly efficient and capital intensive. In the process, prairie agriculture took on an urban character as the farm population declined and local institutions were replaced by those at a distance as producers sought the services available in larger centres. As a result, there has been a wrenching adjustment in the social structure of the prairie region.

The transportation infrastructure, particularly that for grain, required adjustment. But this problem, along with that of elevator capacity, was found to be difficult to resolve. Producer attitudes towards such an adjustment tended to change less quickly than the speed with which they responded economically to their own business environment, even though the two continued to be related.

The situation contributed to the Growing Together exercise, where an action plan was presented having as principles a more market responsive and self-reliant agrifood industry, while also recognizing regional diversity and providing increased environmental sustainability. The development of the prairies has raised issues concerning the sustainability of the type of agriculture adopted. The initial development phase was marked by efforts to cultivate the land and provide an income to support an acceptable lifestyle. Little attention was given to the capability of the land being opened for settlement, soil surveys being non-existent during the heady days of expansion. Concern arose over how best to overcome the recurrent droughts, summer fallow being adopted early in the process.

The impact of the plow and existing summer fallow practices became very evident in the disastrous 1930s. Research, government action and producer ingenuity enabled the problems of the time to be alleviated. While the effects of periods of drought remain, these have been mitigated by the developments in technology. Farming as a result has changed from a way of life to a sophisticated business with repercussions for the rural community, this whole development fostered by government policy designed to reduce risk for the "family farm". However, government intervention with respect to agriculture can be expected to decline as funds become restricted. Greater self-reliance and increased environmental sustainability are projected to become the cornerstones of the new paradigm.

The course of agricultural development on the prairies followed a sequence of models. These include the frontier, the conservation, the urban-industrial impact, the diffusion and the induced innovation models. During the initial period of development on the prairies the frontier model prevailed. Agricultural production increased by expansion of the area cultivated, and thanks to cheaptran sportation Western Europe grew as a market . Essentially, this model prevailed until 1930. The drought years of the 1930s resulted in adoption of the conservation model, where a new form of cultivation practice was found essential. While greater attention was given to sustainability, the replacement of tractor power for horse power rendered some of the attributes of the conservation model inappropriate.

Aspects of the urban-industrial model became apparent in the 1940s as agriculture on the prairies expanded during the second world war. While initially introduced by von Thunen to explain geographical variations in the intensity of farming systems and in the productivity of labour, the rationale of the model later evolved in terms of more effective input and product markets in areas of rapid urban-industrial development. The industrial development of the war years, along with the war itself, stimulated demand for agricultural products. In turn, industrial inputs to enhance agricultural productivity were made available. At the same time a strong off-farm labour market was provided by industry and also by the mobilization of the armed forces, thereby providing a stimulus for higher productivity in agriculture. Regional disparities in productivity and incomes persisted nonetheless.

The diffusion model was in vogue after the end of the war. Agricultural research was making substantial progress in the areas of farm management, varietal development and livestock production. The dissemination of this technical knowledge enhanced productivity on the farm. Producer innovations were refined and tested, improvements in farm equipment being only one example. The applicability of this model effectively ended in the mid-1970s when a major grain and oilseed shortage scare developed which in combination with a rapid increase in oil prices set off a round of inflation.

New cost structures called for cost-saving innovations. The induced innovation model of development was essentially in effect by the 1980s. Technology was developed to facilitate substitution of relatively abundant and cheap inputs of production for those relatively scarce and expensive. This technology was of two types, mechanical, such as tractors and other equipment that allowed the substitution of machines for labour, and biological and chemical, which effectively resulted in "land saving". Machinery enabled each producer to operate a larger area resulting in a decline in labour use per unit of land. Adoption of biological technology facilitated the substitution of labour and other inputs for land. This technology was expressed through new management systems, chemical fertilizers, pesticides and herbicides which when efficiently used resulted in the optimum yield response. Application of science to agriculture has been demonstrated to yield remarkable increases in output. Advances in biotechnology, now an infant science, are forecast to give rise to a transition in farm practices comparable to the shift from horse power to tractor power. The induced innovation model also contains the elements of induced institutional innovation.

Researchers respond to the changes in the economic environment within which they work. Their rewards lie in the material benefits received for their work and, perhaps even more importantly, in the satisfaction attached to overcoming constraints on output. Technological change in turn gives rise to institutional change. The need for viable institutions capable of supporting more rapid agricultural growth and rural development becomes compelling. The implication is that a growing interdependence becomes essential between the natural and social sciences due to the impact of technology on the rural development process. Similar innovation in rural institutions become required if the rural development process is to be fully effective.

The induced innovation model continues to be applicable to agricultural development on the prairies. Technology, however, with its impact on the size of farm, has a major impact on rural communities. Those which survive are taking on more of the characteristics of an urban community. To succeed, community development programs must therefore address both rural and urban concerns. Such concerns include the quality of education, health facilities, housing and other essential services. A major challenge arising for rural communities as a result of induced technology becomes how the opportunities and challenges for growth can be met without losing the attractions which have led farm people and other rural residents to prefer rural life. Agriculture continues to be faced with many challenges. One is the need to make farming economically viable, to provide the caretakers with the means and the incentives to conserve two of the nation's most important resources, the soil and water.

Another is to have institutional and scientific support services, which combine far-sighted wisdom with immediate practical value. Given these, agriculture will continue to provide food for the nation and the world at large. The history of agricultural development on the plains therefore provides a proper introduction to the historical meaning of sustainable agriculture in the Plains region. The sustainability of agriculture was a very essential objective in the development of the region. Vagaries of the climate, the lack of infrastructure and the need for new technology all had to be overcome. Existing forms of agriculture reflect development in the past as modified by the policies of the present. The current emphasis on sustainability therefore represents only one position along a development course which started in the 1880s. It is nonetheless recognized that the present concept of sustainability in agriculture is much more comprehensive than was originally the case on the Plains.

A number of issues concerning sustainable agriculture and resource use on the Great Plains have been identified. They range from the measurement of sustainability to the potential impacts of global change. The issues identified below do not constitute a comprehensive list but do represent those considered to be of major importance. They are also indicative of the diversity of the issues. These issues are not necessarily independent.

Measurement of Sustainability

While some progress has been made in assessing the sustainability of particular production processes, much remains to be accomplished. Measurement is complicated by the need to establish an appropriate time frame. The tendency is to look at an infinitely long period. This approach is not necessarily appropriate since it is impossible to accurately predict future events. The concern is with the capacity of natural resources in combination with physical and biological systems to provide the desired flow of output in the future. A further complication is that other factors impinge on production. Investments in intellectual and physical capital result in development of new or improved production practices which may involve using other resources to produce the output desired. An example is the substitution of nuclear energy for coal in the production of electrical power.

There are two measures of sustainability to be considered: the economic and the physical. Richard Gray among others has devoted considerable efforts to development of economic measures, while the work of Zentner provides an insight into determination of the physical as related to the soil resource used in agriculture.

An economic measurement of sustainability is difficult, as there are several viewpoints to consider. Gray stated that two general views prevail concerning sustainability in agriculture, one by consumers and one by producers. Consumers view sustainability in terms of its capacity to provide an a bundance of quality food. Producers view sustainability as an income generating activity with economic and social value. The concern they have is with maintaining a net return from the sale of agricultural products.

In economic terms, consumers assess sustainability in terms of maintaining a level of consumer surplus in the consumption of food over time. On the other hand, in economic terms, producers regard sustainability as the maintenance of a producer surplus or economic rent over time. This dichotomy of interest provides a basis of conflict and attracts attention to the difficulty experienced when trying to measure sustainability. From the consumer's standpoint, security of the food supply is upper most particularly among those who have previously experienced shortages or indeed famine.

Measurement of the value of food security becomes almost impossible. Time preference becomes a complicating factor as does trying to aggregate the preferences of individuals. Particular issues that are raised when measuring sustainability are the discount rate; private versus social costs (negative externalities); non - market benefits (positive externalities); economic flexibility; and income/risk preferences. In addition to maintaining a producer surplus is the need to sustain farm families. Furthermore, sustainability must be measured taking into account uncertainty. Gray suggests that the elasticity of supply (the change in output relative to a change in price) may be a relevant indicator of sustainability. Output in response to price changes is impacted by technologies and available inputs.

Consequently, society should support policies which enhance technology through research. Since the tools of government policy impinge on sustainability and policy results from the political process, economic measurement of sustainability becomes a very difficult and hazardous task. Receipts from government payments, these not being sustainable, ranged from 52 to 191 percent of farm net income during the period. The range was even greater for individual provinces, Saskatchewan being a prime example where government payments as a proportion of net income ranged from 56 percent to 701 percent. The returns from crop sales were greatly depressed over the period due to the export subsidies provided by other grain exporters, the European Economic Community (EEC) and the United States (U.S.) in particular. While the proportional contribution of agriculture to the Gross Domestic Product (GDP) has declined over time it remains significant in the prairie provinces.

They also confirm that severe problems exist in the economic sustainability of agriculture on the prairies. There are indications that the potential exists for increasing aggregate farm income by an expansion of livestock output, an activity which holds forth the prospect of also increasing the sustainability of agriculture in physical terms. In contrast to the economic measurement of sustainability, a physical measurement is relatively straightforward. It deals with particular processes or uses of resources. Zentner illustrates such measurement with reference to the soil resource. He notes that since the soil resource used in agriculture is largely privately owned, producers can be expected to organize their activities in a manner that maximizes their private benefits.

On the other hand, society desires to maximize the social benefits. There is a common interest in conserving the soil since failure to do so increases the marginal costs of production and reduces the future streams of private and social benefits. However, several reasons exist as to why producers adopt a set of soil use strategies which differ from those held desirable. They may lack information on the depletive consequences of particular production practices; they may be faced with constraints irrelevant to society as a whole, including the need to accept lower short run profits as a result of soil conservation practices, institutional restrictions such as quota systems and government policies which create distortions and encourage the use of soil depletive practices; they may have different planning horizons and time preferences; and they may have a short term view on externalities such as saline seeps, soil erosion and contamination of ground water.

Three factors must be considered when estimating the long term consequences of soil degradation. These factors are the value of agricultural production that is permanently lost or not forthcoming as a result of reduced soil productivity; the cost of the additional inputs necessary for agricultural production; and the social costs of the externalities being generated. Accurate measurement of these factors is an exceedingly complex process. Furthermore, estimation of the long term effects on soil productivity is subject to substantial error. In addition, establishment of the economic consequences of soil degradation involves forecasts of future reductions in productivity while recognizing the possibility of development of new technologies.

Finally, the economic consequences of soil degradation alone can be determined by estimating the returns otherwise received from the lost production. It should be noted that the additional output forthcoming from conservation will have a negative impact on price. Evaluation of sustainability, even in the case of a particular resource used in agriculture, is replete with problems. A number of value judgments are required since some of the variables do not lend themselves to objective measurement. Technologies which tend to conceal the effects of degradation in terms of yield continue to be developed. Consequently, measurement of the sustainability of a resource, such as the soil, is difficult over an extended period of time. On the other hand, exact measures may not necessarily be mandatory when attempting to evaluate sustainability since in many situations the degradation is obvious and, as in the case of the soil, should stimulate action to adopt conservation practices.

Land Use

The soils of the prairies developed under grass land vegetation which supported substantial populations of numerous animal species. At the same time, sloughs and small lakes dotted the landscape providing habitat for birds and other wildlife. Trees in river valleys also provided sanctuary. The vegetation of the region varied from short grasses in the drier sections to tall grasses in areas of increased rainfall to reeds and sedges in low lying areas. Numerous species of plants native to the prairies also existed. The prairie landscape prior to 1870 exhibited considerable diversity.

The advent of settlement on the prairies substantially modified the prairie landscape as large areas of the grasslands were put to the plow. In the process, decimation of the previous large herds of animals, particularly the bison, took place along with their dependent predators and scavengers. The pressure to expand the cultivated area gave rise to drainage of wetlands thereby further destroying habitat for wildlife. Over time a grain monoculture developed in much of the prairie area, a farming practice encouraged by prevailing government policies. It therefore becomes apparent that land use on the prairies has evolved over time.

Land has multiple uses as was demonstrated during the development phase. Prime uses include animal and plant habitat and agriculture. As the population expands there is greater demand for land for recreational purposes and for housing in urban centres. With the primary use of the land being for grain production, the incipient decline in the grain economy gives rise to the issue of whether the present pattern of land use is appropriate. The relative contribution of agriculture to the economy generally has also declined, further compounding the land use issue. On the other hand, demand for land for purposes other than agriculture is increasing.

While the trends in land use are similar for each of the prairie provinces, the degree of change differs. The area of land under crops has risen over time. The additional area has been made possible largely as a result of a decline in land devoted to summer fallow and in the case of Saskatchewan and Alberta to an increase in the area of improved land. Over the ten year period, 1981-1991, the area in summer fallow in Manitoba declined by 50 percent, in Saskatchewan by 16 percent, and in Alberta by 20 percent.

While data for the area in improved land are not available for 1991, between 1981 and 1985 a decline of 2 percent is indicated for Manitoba while increases of 2 percent and 3 percent were registered for Saskatchewan and Alberta, these increases being stimulated by the agricultural policies in effect. In the process, there has been a reduction in the area of wetlands on the prairies as indicated by Caswell. His data are illustrated in Figure 1. Current land use on the prairies has reduced the quality and quantity of diverse wildlife habitat in Canada. While farmlands and rangelands do provide enhanced habitat for certain species, other species have declined as a result of agricultural expansion and production practices.

Other uses for land are not necessarily competitive with agriculture. While the land is a critical resource for agriculture, it is also a critical resource for the environment and for society. Restoration of some of the land as habitat for wildlife can be complementary to agriculture. At the same time, diversion of some of the land presently used for grain production into the production of livestock can result in agriculture being more sustainable.

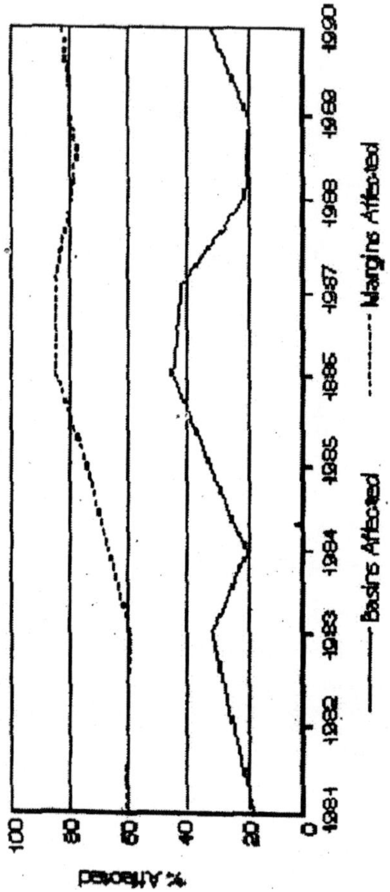

Figure 1.

Percentage of Prairie Ponds with Margins or Basins Affected by Agricultural Practices, 1981-90

The purposes for which land is used reflect their relative profitability. The profitability of a particular use in turn continues to reflect existing policies. Notwithstanding this situation, the demands of society are exerting pressure on the current pattern of land use. Eventually these demands will result in the land less suitable for grain production reverting to uses similar to those prevailing before the time of settlement.

Soil is degraded as a result of processes that reduce its productivity. Such processes usually arise from poor management of the resource. Soils are degraded primarily by erosion, loss of organic matter, salinization and acidification. The first two processes are interrelated. Acidification is not a major problem except in areas where cultivation has been extended into previously forested areas. Salinization is more prevalent.

Soil erosion

Soil erosion reduces soil productivity through losses of nutrients, water storage capacity and organic matter. The losses in terms of productive capacity are significant though presently obscured by the application of fertilizer. Annual top soil losses in terms of lost productivity on the prairies are estimated to be in the millions of dollars. The present value of the accumulated annual losses is measured in hundreds of millions of dollars, the actual losses being difficult to quantify.

Erosion has taken place over the history of cultivation on the prairies. Both wind and water erosion result in the removal of the finer soil particles, thus leading to compaction of the soil and poor soiltilth. Water erosion is a significant problem on long sloping fields, particularly those summer fallowed or left free of cover. The valuable top soil removed is deposited in depressions, road ditches and waterways giving rise to additional costs. Wind erosion is a major problem when there is little cover by crops or residues as in the spring months particularly during dry seasons. Not only are valuable nutrients lost but the soil deposited elsewhere becomes a problem. Indeed, seedling crops may be destroyed.

Table 1.

Trends in Land Use, Prairie Provinces, 1981-1991

	1981	*1986*	*1991*
		thousand hectares	
Manitoba			
Area in Farms	7,615	7,740	7,725
Under Crops	4,420	4,519	4,761
Improved Pasture	353	275	341
Summer fallow	598	509	297
All Other Land	2,245	2,437	2,326
Saskatche wan			
Area in Farms	25,947	26,599	26,865
Under Crops	11,741	13,326	13,459
Improved Pasture	975	879	1,076
Summer fallow	6,704	5,658	5,713
All Other Land	6,526	6,738	6,618
Alberta			
Area in Farms	19,109	20,655	20,811
Under Crops	8,441	9,163	9,292
Improved Pasture	1,581	1,377	1,742
Summer fallow	2,205	2,127	1,771
All Other Land	6,880	7,989	8,005

Sparrow has expressed annual soil losses in different terms. He estimated the annual soil loss on the prairies by wind to be about 160 million tonnes and that by water to be 117 million tonnes, these losses from erosion greatly exceeding the rate of soil formation. The annual loss in potential grain production arising from soil erosion during the period of cultivation he considered to be 171 million bushels or 4.6 million tonnes of wheat. He argued that 15 percent of the lost production could not be recovered by additions of fertilizer and estimated the value of lost production at $129 million annually, lending credence to the title of his report Soil at Risk:

The physical effects of erosion are painfully evident on the knolls and sloping segments of the prairie landscape. It has led to a gradual evolution of farming practices from an initial use of the plow, through use of blade type cultivation to use of minimum tillage. This progression has enabled increasing amounts of trash cover to be maintained over time, thereby reducing erosion. Moreover, in more recent years the width of fields has increased as a reflecting the size of the equipment used, which resulted from the economies of size in farm operations. In addition, intensive tillage operations to incorporate herbicides and fertilizers were adopted. These latter events work against the soil conservation otherwise achieved through erosion maintenance of trash cover.

Fortunately, a reduction in the cost of suitable chemicals has made the use of chemical fallow more economical, thereby enabling adoption of minimum tillage practices. This change in practice has been driven by the need to minimize production costs, the protection of the land from erosion being a complementary benefit. At the same time, summer fallowing has largely become confined to the brown soil zone as rotations have been extended in the dark brown zone and continuous cropping has become general in the black soil zone. Losses from soil erosion can therefore be expected to be less extensive in the future than in the past.

The main chemical constituents of soil organic matter are carbon and nitrogen. The former is the energy source for most soil microbes while the latter is one of the most important nutrients for plants. Other necessary components of organic matter are phosphorous and sulfur. Under the ecosystem prevailing previous to cultivation, mineralization and immobilization processes were closely integrated. Plant growth (immobilization) and mineralization of soil organic matter occurred simultaneously. Therefore losses of nutrients were minimized allowing for the accretion of organic matter over time. Carbon transfer through the soil is basic to the functioning of the soil-plant system.

Primary productivity determines the amount of carbon entering the system while the amount leaving is controlled by soil micro-organisms. In natural ecosystems, the amount of carbon transferred by means of both processes appears equal. Upon cultivation of the soil for the production of annual crops, productivity and mineralization are uncoupled, particularly when the fields are bare of vegetation. Total inputs of carbon to the soil are altered due to differences both in plant characteristics and to allocation of carbon for above and below ground production.

The practice of summer fallowing has hastened the loss of soil organic matter. Frequent tillage during summer fallowing has increased the rate of decomposition of organic matter while at the same time disrupting soil aggregation and increasing its susceptibility to erosion. There has also been a loss of the more readily available nitrogen.

The magnitude of the losses of organic matter under cereal production has been established as being in the range of 20 to 50 percent in the cultivated layer according to the soil classification. Some estimated in 1981 that organic matter losses in the topmost layer of soil averaged 41 percent in the brown soil zone, 44 percent in the dark brown soil zone, 49 percent in the black soil zone and 36 percent in the grey soil zone. Research has established that by improved cultural practices, the decline in organic matter can be halted and indeed reversed. Nonetheless, additional nitrogen is now required to maintain yields on many prairie soils. Regardless of the productivity losses identified, annual wheat yields after an initial decline following settlement have trended upward over the last 50 years. This trend is illustrated in Figure 2.

The increase in yield identified by Hedlinreflects the impact of adopting higher levels of technology, one of these being the application of fertilizer. The aggregate quantities of nitrogen and phosphorous now being applied is approximately that used by the crops. Uncertainty remains, however, as to how soils will behave under such changed cultural practices as snow trapping, minimum tillage, green manuring and chemical follow. This uncertainty arises because soil changes occur slowly and the effects of agronomic practices take years to become evident.

Soil Salinity

Even before cultivation, large areas of the prairies contained saline soils. However, an expansion of these areas has occurred under cultivation. The rate of increase in the salinized area is related in part to the increase in ground water over time.

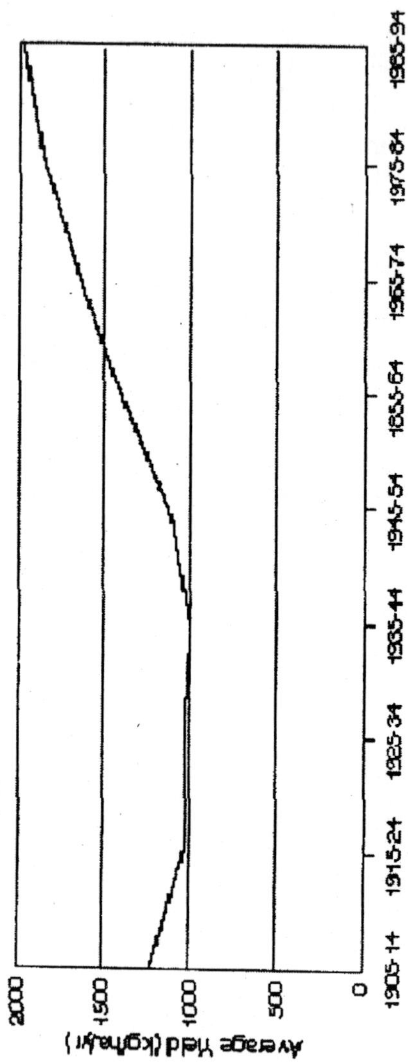

Figure 2

Trends in Wheat Yields in the Prairie Provinces, 1905-84

Salinization occurs as water containing soluble salts moves upward through the soil horizon. Evapotranspiration concentrates the salts in the soil solution to reach levels in the root zone that are detrimental to crop growth. Any technique which removes water from the subsurface such as drainage, use of deep rooted perennials, reducing summer fallow or continuous cropping diminishes the danger of salinization. Once salinization has occurred, the remedy is to leach the salts out of the root zone. The most viable way to accomplish this cleansing is to use all available precipitation to grow crops and thus reduce the accumulation of ground water. This indicates the desirability of reducing the acreage in summer fallow on the prairies. Such a reduction would also lead to a significant decline in the leaching of nitrogen into the subsoil.

Salinization of dryland agricultural soil constitutes a major problem in southern and central regions of Alberta and Saskatchewan and in a few scattered areas of Manitoba. Approximately 2.2 million hectares of improved land are considered to be subject to secondary salinity. The economic impact of this salinity is said to be in the range of $104 - $257 million annually. Acidification is not a major problem in other than a limited part of the prairies.

Acidification can be overcome by application of lime. However, heavy applications of nitrogen fertilizers over time have been found to result in soilac idification. Consequently, acidification may become a significant problem in parts of the prairie region under intensive cropping practices which require major inputs of nitrogen fertilizer.

While degradation of the soil resource can be reduced or halted by changes in cultural practices, unless these are demonstrated to be economic they are unlikely to be adopted even though desirable from the standpoint of society. Producers are more likely to be interested in maximizing profits in the short - run rather than be concerned with conservation of natural resources and protection of the environment. Any sizable shift in cropping practices by producers will depend on their ability to survive difficult economic times.

Preservation of Biodiversity

Biodiversity is defined as "the variability among living organisms from all sources including, inter alia, terrestrial, marine and other aquatic systems and the ecological complexes of which they are a part; this includes diversity within species, between species and of ecosystems." Biodiversity is considered essential for the resilience of ecosystems. Many ecosystem processes regulate conditions for life. Holling argues that this role provides the rationale for giving priority to rehabilitation, protection of ecosystems and land use management. Resilience enables ecosystems to return to a steady state after being subjected to an unusual event denoted as being a "surprise". The ecosystems may not necessarily return to the same state as before. In other words, ecosystems are subject to discontinuous change. However, there remains the risk that intervention in ecosystems by humans can result in a reduction of their resilient nature rendering them incapable of dealing adequately with unusual events. Holling describes the situation with respect to ecosystems as follows:

"Ecosystems have a natural rhythm of change, the amplitude and frequency of which is determined by internal processes and structures in response to past external variabilities. These rhythms alternate periods of increasing organization and stasis with periods of reorganization and renewal. They determine the degree of productivity and resilience of ecosystems. Modern technological man affects these patterns and their causes in two ways.

First, traditional resource management institutions constrain rhythms by restricting them temporally and homogenizing them spatially. Internal biophysical relationships then change, leading to systems of increasing fragility, i.e. to a reduced resilience. Moreover, modern man and his institutions operate with a different historical rhythm which can mask indications of slowly increasing fragility and can inhibit effective adaptive responses, resulting in the increased likelihood of internally generated surprises, i.e. crises.

Second, the increasing extent and intensity of modern industrial and agricultural activities have modified and accelerated many global atmospheric processes, thereby changing the external variability experienced by ecosystems. This imposes another set of adaptive pressures on ecosystems when they are already subject to local ones. As a consequence, locally generated surprises can be more frequently affected by global phenomena, and in turn can affect these global phenomena in a web of global ecological dependencies." Loss of biodiversity has profound implications for development. Biological resources are renewable and output can be increased under appropriate management. Natural habitats which can maintain productivity without significant management have the ability to provide means for human survival. Highly diverse natural ecosystems also provide important ecological services. These include maintaining hydrological cycles, regulating climate, contributing to the process of soil formation and maturation, storing and cycling essential nutrients, absorbing and breaking down pollutants and providing sites for inspiration, tourism, recreation and research.

While nature has some built-in redundancy few data are available on which species are particularly important in the functioning of ecosystems so that the actual value of specific losses of biodiversity is not available. There is need, however, to at least maintain biodiversity by maximizing the number of different species conserved in sufficient quantities to assure survival. The many values of biodiversity and its importance for development suggest why biodiversity conservation differs from traditional nature conservation.

Biodiversity conservation entails a shift from a defensive posture to an offensive effort seeking to meet peoples' needs from biological resources while ensuring the long-term sustainability of earth's biotic wealth. It involves not only the protection of wild species, but also the safeguarding of the genetic diversity of cultivated and domestic species and their wild relatives. This goal applies to modified and intensively managed ecosystems as well as natural ones and is pursued in the human interest and for the human benefit. Conservation of biodiversity seeks to maintain the human life support system provided by nature and the living resources essential for development. Biological resources support development in virtually all sectors and affect those who live in cities as well as those in the country.

Agriculture under good management is an example of the management of a modified ecosystem to yield what is perceived to be optimal productivity. The system is inevitably somewhat impoverished as predators and competitors are eliminated or reduced and the population structure is altered in order to enhance yields. Yet it is ecologically sound and essential to human welfare. Such a system both affects and depends on the more natural ecosystems.

Biodiversity may, therefore, be seen as an indicator of environmental health. Success in maintaining biodiversity must take into account both spatial and temporal factors. It is not, however, possible to ensure a constant level of biodiversity at a particular location over time. Agriculture by its close attachment to the land is in a position to greatly influence biological diversity on the prairies. Agriculture has substantially changed the prairie landscape. Annual crops have replaced the grasses which previously covered the land. In fact, modern agriculture can be said to have had a greater impact on biodiversity in the Plains region than any other activity.

As agriculture became more intensive, wetlands were drained, trees removed and grasslands cultivated. These actions, while being advantageous for the use of large modern equipment, effectively destroyed the habitat for many plant and animal species. In no area was this development more effective than on the open prairie. These actions were largely induced by public policy which directly or indirectly fostered expansion of the cultivated area. Indeed, more than 80 percent of the prairie landscape has been altered by agriculture. Drainage for use in agriculture has been responsible for 85 percent of the total land drained.

The impact on the bird population, because of following this course, has been significant since the sloughs and small lakes drained provided the habitat for more than 60 percent of the waterfowl on the continent. It should therefore not be surprising that waterfowl populations have declined over time. Continued expansion of the area in agriculture, whether by removal of indigenous prairie trees or extension into the parkland area, serves to reduce the habitat for wildlife.

Animals also suffer from the reduction in habitat, often being forced to retreat into more remote areas. In many parts of the prairies native vegetation has become a thing of the past, stimulating organizations to strive to protect any areas remaining. The extent to which agriculture on the prairies has changed the landscape has been described by Gauthier and Henry. Only one percent of the tall grass prairie, 18 percent of the short-grass prairie, 25 percent of the mixed grass prairie and 25 percent of the aspen parkland remain. Furthermore, conversion of these lands to agriculture continues, as a result largely of agricultural policy.

The remaining areas of untouched wildlife habitat have become progressively more fragmented, more isolated, and often too small to sustain viable populations of species that had once been abundant. In no case is this predicament more evident than in that of wetlands, which are said to be disappearing at the rate of 0.5 percent annually in Alberta.

On the prairies, 60 percent of the wetland basins and 80 percent of the habitat surrounding these basins are affected by farming each year. This occurrence has a significant impact on waterfowl. Grazing by cattle and other domestic animals reduces plant cover and the supply of food and shelter for meadow and grassland species of mammals, birds and invertebrates. On the other hand, some forms of wildlife thrive under the conditions arising from agriculture. Farm shelterbelts provide cover for numerous species of birds. Other species have benefited from the increased feeding opportunities provided by agricultural crops. These include ducks, geese, sandhill cranes, robins and blackbirds. However, biodiversity remains threatened by agricultural operations, as 12 of the 62 species of plants considered endangered are wetland species. Furthermore, diseases of farm livestock have been known to be transmitted to native animals with devastating results. Conversely, diseases can be transmitted from native animals to those on the farm. Maintenance of the current level of biodiversity on the prairies, which reflects the impact of agriculture on habitat, cannot be considered particularly difficult .

Biodiversity in terms of numbers of different species on the prairies has not been so much affected as have the numbers remaining within given species. The factors contributing to the decline in numbers include agricultural policies, regional development plans, institutional structures, world trade, characteristics unique to the rural economy and the world view prevailing in rural areas.

Water use and Quality on the Prairies

The great need for water by agriculture is one of the reasons why water is in short supply on the prairies. The lack of water in the southern prairies contributed to the devastating experience of farmers in the drought area of the 1930s. Indeed, during the 1961 - 90 period, the average annual precipitation in Regina was 362 mm while the average evaporation rate was 1016 mm, resulting in a net deficit of 654 mm. This difference illustrates the need to conserve moisture supplies and reduce the evaporation rate through the application of technology. Furthermore, quality water constitutes only part of the supply available. On the other hand, agricultural operations can contribute to loss of quality in water. A further complicating factor is the unequal distribution of water in the prairie region. Consequently, water use and quality have become a major issue on the prairies with regard to the sustainability of agriculture in the region.

The low level of precipitation and its uneven distribution on the prairies in combination with evapotranspiration has made redistribution of the available supply of water of paramount importance. Redistribution may be accomplished by on-farm techniques to retain the precipitation that falls. More generally, water is drained from undesirable locations to others more desirable as well as transferred within basins and also between basins. The latter techniques decrease the impact of the variability of precipitation. Such redistribution activities are not without cost and may have an undesirable effect on the environment.

While ground water is the most abundant and widely distributed water source, the water quality, yield and dependability of the aquifers can vary considerably. Deep aquifers, while more resistant to drought, tend to have high treatment costs in order to make them potable. One of the first activities of PFRA following its inception in 1935 was to embark on construction of on-farm dugouts to capture any runoff for domestic and livestock use. Dugouts have been termed as an inelegant a collection of holes as ever designed by man. They did nothing aesthetically for the landscape. They were utilitarian in the extreme. Yet, if acceptance is a measure of worth, few more valuable projects were ever devised . PFRA projects also included stock watering dams and other on farm water supply projects. To date, over 200,000 of such on-farm water projects have been completed through joint producer and PFRA action.

Comprehensive planning combined with coordination and sound water management are fundamental requirements for the development and maintenance of secure water supplies. Increased water use combined with occasional droughts have resulted in the failure of many surface sources and shallow water aquifers to provide sufficient water for farms and rural communities. As a result, within the prairies with the cooperation of PFRA virtually every major system has been modified by means of reservoirs so that the flow of water in peak times is made available for use in drought periods. Of the approximately 770 dams in the prairie provinces, 53 percent are in Saskatchewan, 40 percent in Alberta and 7 percent in Manitoba. The most monumental of these river projects is the Gardiner Dam on the Saskatchewan River at Outlook, Saskatchewan, which created Lake Diefenbaker. Even though the irrigation potential remains largely undeveloped, the benefit to cost ratio of this multi-use project generally exceeds one without the uncalculated benefit of the climate modification associated with such a large body of water. Originally proposed by Henry Hind in 1859, the dam was completed in 1968.

Irrigation has been shown to enable a form of sustainable agriculture to develop in an arid region, with a complementary output of products which can provide the basis for development of substantial value added industries. Economic stability is achieved following the development of a successful irrigation area. These benefits serve to offset public investments in the associated water management projects. It should be noted that the development of irrigation in Southern Alberta early in the 1900s enabled the area to escape the disastrous consequences of the "dust bowl" which developed further east on the prairies during the 1930s.

Irrigation accounts for approximately 88 percent of the total water used for agriculture. Irrigation is practised on about one million hectares of land in the brown and dark brown soil zones, the crops produced being primarily cereal grains. Recent droughts have stimulated interest in additional water management projects. Unless the water is wisely used soil salinity problems may be experienced. At the same time inter basin transfers can give rise to introduction of new species and to habitat loss for fish and wildlife.

Redistribution of water tends to be expensive. Use of public funds for this purpose is subject to increasing scrutiny. Questions arise as to whether use of these funds would return greater benefits if spent for other purposes. The net pay-off of most past irrigation projects has been positive. Many projects are multipurpose in nature providing water supplies for power generation, municipal use, recreation, and wildlife habitat as well as for agriculture.

Not only does the sustainability of agriculture have a water quantity requirement but also one of water quality. For human and livestock use, the quality requirements are quite stringent. For crops, levels of tolerance are somewhat higher. Even treated effluent from municipal sewage plants can be used to irrigate crops.

Agriculture itself can become a source of water pollution both on and off location. Local pollution sources include processing plants, and confined livestock operations. The primary pollutants are phosphorous, nitrogen and organic materials. Off-site pollution from agriculture includes phosphorous, nitrogen, pesticides and sedimentation of streams and lakes.

Water has and no doubt will continue to have a major impact on the sustainability of agriculture on the prairies. Snow management can add three centimeters of water for the production of the next crop. The additional water is of great value, particularly in the brown and dark brown soil zones where on average the amount available is inadequate. The additional water from snow management can often provide an alternative to summer fallow as a means to increase soil moisture. Adoption of water efficient cropping systems such as minimum tillage or zero tillage can also alleviate the moisture shortage in crop production. Control of weeds and use of fertilizer improve the efficiency with which the available moisture is used in the production of crops. All these activities point to the desirability of improved water management on the farm. Such management could overcome some of the need to summer fallow in the brown and dark brown soil zones for the purpose of storing water for the next crop.

It must be emphasized that sustainable agriculture on the prairies depends on an adequate supply of water. In the past, there was little outside competition for the available supply. However, competition is expected to increase as municipal demands rise along with the population and as demands from industry expand. In consequence, efficiency in the use of water will become mandatory. In this environment, water from off-farm sources can be expected to be only available at full cost, rendering most proposed additional water storage works for agricultural purposes uneconomical. In addition, greater scrutiny of the pollution of water by agricultural operations can be expected. On the other hand, increasing concern for the maintenance of wetlands for environmental purposes should promote the sustainability of agriculture.

Use of Common Property

In agricultural production a number of so-called "free" goods are used as inputs, primary examples being air and water. In the process, both negative and positive effects may become evident. Misuse or abuse of these goods may arise from the fact they are "free". In terms of environmental impact all goods entail a cost. Indeed, the local landscape provides public goods including amenities such as a satisfying setting, wildlife, air and water. At the same time, private goods such as crops and livestock are also produced. The impact of agriculture on the use of public goods gives rise to tension between agricultural producers and other sectors of society. Producers follow their private economic interests in management of their farms and may in the process reduce, eliminate or impair the public goods desired by others.

Unless values are attached to both the public and private goods along with those arising from their impairment or enhancement, the true cost of production will not be taken into account: "There has, in other words, to be a trade-off between what are normally understood as 'economic goods', such as food and chargeable rural leisure services, and the sustenance of less measurable environmental goods. The less measurable or definable services have to be properly valued. Society must choose on the basis of a full understanding of the choice in question. That means that the economic value of the environmental cost, if one is to be incurred, must be understood."

During the process of production, land may be degraded as a result of a producer's actions which may be encouraged by the instruments of government policy. Habitat may be destroyed in the process. This illustrates the need for soil conservation as degradation affects both the producer and those using the services produced. Often, the producer is unable to capture all the benefits of conservation. Determination of how the benefits may be achieved therefore presents a problem even though the results of conservation are very apparent in purer air, less sedimentation in streams, greater wildlife habitat and other benefits.

The above situation has led to a search for effective means by which the desired end may be achieved. Such means include government regulation, purchase of property *rights, financial incentives or fines, cross compliance in* government policies, and moral suasion. However, much attention has also been given to valuation of the benefits and the costs to the environment as well as to the returns from production. Trade -offs between aspects of these environmental services will undoubtedly have to be made. Non-market evaluation provides one form for evaluating the magnitude of these trade-offs. Also, some insight into the relevant policy approach has also been derived from evidence supported by nonmarket evaluation. The concern over off-site impacts of agriculture implies an externality control policy may be more relevant than on-site productivity policies." When the market fails to establish value in terms of price, i.e., market failure, there is the potential to have an unsustainable system. Those who pay the full costs of production sometimes do not receive the benefits and conversely sometimes the amount paid does not cover the full costs of production. In agriculture, producers may not be compensated for soil conservation and therefore may not engage in the practice. As a result, for example, sloughs may be drained which remove habitat for wildlife. Thus many air, soil and water quality problems can be traced to market failure.

Where producers harm the environment by contamination of air or water, regulation appears appropriate since the perpetrators of the problem may not easily be identified, rendering enforcement of the "polluter pays principle" infeasible. Tension between agricultural producers and society in the use of the environment will increase unless the common property issue is resolved.

Common property implies that goods are essentially available for the taking. No property rights are involved. A frequent prescription for overcoming the tendency to abuse public goods is privatization. This principle has application to agriculture with respect to land tenure. A substantial proportion of the land on the prairies is operated by tenants who hold either cash or share leases. Unless the person operating the land is credited with the value of any improvements made, there is little incentive to invest in long term measures such as soil conservation. When land prices rise the tenant may be unable to resist a change in the ownership of the land, some purchasers merely buying the land for speculative purposes or as a hedge against inflation. These individuals have little interest in conservation. The land also may not be sold on the open market and therefore not purchased for its highest production value. This may or may not be synonymous with environmentally benign land use. Furthermore, the tenant has less collateral to use as the base for borrowing funds, the purpose of which would be improving the productivity of the land.

Private property does not resolve the issue of external effects imposed by one owner on another unless both parties can negotiate the externality. In order to achieve a situation conducive to soil conservation it becomes essential that the responsibilities of the tenant and the landlord be carefully spelled out with the agreement covering an extended period of time. The present pattern of land tenure on the prairies may therefore have implications for maintenance of the soil resource and ultimately on the sustainability of agriculture. As inferred by Pearce and Warford, absence of operator ownership has degradation.

The proportion of the land operated by those renting or holding leases rose from 31 to 39 percent. Over the five year period 1986 to 1991, the respective proportions remained relatively stable. Nonetheless, tenancy remains an issue holding forth the prospect of affecting the sustainability of agriculture on the prairies.

Economics plays a major role in the sustainability of agriculture. To be sustainable, agriculture must be economically viable. That is not to imply that all producers or all forms of production must be viable in any region at a particular time. Variations in the returns are necessary and indeed desirable. They stimulate change in the mix of products produced, thereby achieving consistency with the demands of the market.

The emphasis on grain production on the prairies combined with the associated dependence on export markets has rendered the prairie region highly susceptible to changes in the world economy. Trade policies of other nations can and do impact severely on the region. In consequence, government has been moved to alleviate the ensuing distress through various programs of assistance. Farm income received from the marketplace has varied significantly over time, a pattern no more evident in any combination of products than in grains and oilseeds. In recent years, shortfalls of producer income have been partly assuaged by large infusions of government funds.

The magnitude of these infusions is very impressive. Total federal government assistance programs for the grains and oilseeds sector reached $3,923 million in 1991/92. As a result, the fluctuations in farm income were much less than they otherwise would have been, thereby reducing the financial stress of many producers. While the money was very acceptable on the part of the recipients, little was accomplished in terms of reorienting of agriculture on the prairies in a more sustainable direction. Land prices were maintained at levels higher than those which could be supported by returns from the market. Desirable change in farm structure was retarded and little change occurred in the mix of products produced.

A propensity for receipts from the six major grains to decline is evident during the 1987-91 period, a decline of 12 percent being recorded. Meanwhile receipts from other crops increased by 12 percent. Receipts from sales of livestock and livestock products increased by 8 percent. During the same period, program payments declined by 47 percent. The net result was that total receipts declined by 4 percent. The program payments while enhancing income stability on the prairies also can be said to have limited the urgency for needed farm adjustment.

By providing subsidies to producers to reduce food costs, consumers are limited in their ability to choose where they want to spend their money. Consumers are essentially supplying the producers with subsidies for food production. Yet it is possible that consumers are more interested in spending their money on products which provide them with greater satisfaction. The government subsidies force all taxpayers to assist in lowering food costs. When all externalities are accounted for in the price, consumers who would rather spend their money on something else are made worse off, as are those who would spend their money on other products that have lower production costs. Moreover, the viability of those businesses where the consumer's funds would otherwise have been spent comes into question.

Government infusion of funds to agriculture cannot be expected to continue as it has because deficits have accumulated to the extent that any additional debt is difficult to support. Under this situation it becomes necessary to rely on the marketplace for returns, rendering an even more satisfying resolution, at least in part, to the trade issue through the new GATT agreement. A more diverse agriculture on the prairies appears desirable , particularly between farms. Unfortunately, transport policy runs counter to indications from the marketplace that a more livestock oriented farm economy would provide increased returns and absorb greater quantities of the grain produced. Such an economy would enhance the sustainability of agriculture.

While it has been argued that higher prices for products would overcome the economic problems of agriculture, history has shown that this is not necessarily the case. Changes in farm structure occur in high price periods as producers in favourable financial circumstances, or those willing to assume high levels of debt bid up land prices. In low price periods, structural change continues as the farms of those who leave the industry are purchased by those having more capital. Regardless of government assistance policies, structural change continues.

On the prairies, this structural change is driven by changing technology in grain production. This technology entails minimum tillage practices that utilize improved cultivation and seeding equipment, materials which are economically feasible only when used on large acreages. While this technology is positive for sustainability, substantial structural change is occurring that is rendering many farm workers surplus. Where farms specialize in grain production, significant underemployment in farm activities occurs during a substantial portion of the year.

Under existing economic conditions, additional income is needed to support the farm and farm family. Off-farm employment is often undertaken, initially as a means to overcome a short term economic crunch. Later, off-farm employment becomes a deliberate choice to diversify and raise the level of family income in order to enjoy the amenities available in urban areas. While technology in agriculture has enabled farm men to enter the off-farm labour force, participation by farm women has been rendered feasible by labour saving devices in the home and a reduction in family size. A substantial proportion of farm women are therefore able to capitalize on their training by obtaining employment at the professional and managerial levels in nearby communities. Improve d transportation facilities have rendered feasible greater commuting distances from the farm to the employment site.

Off-farm employment may be said to represent a partial adjustment of labour away from agriculture while continuing farm production. This measure enables the attributes of rural living to continue to be enjoyed. Over time the returns from off-farm employment may outweigh those received from agriculture.

Off-farm employment income has become a substantial proportion of the total income received by farm families. While off-farm income represents a proportionately larger share of total income for families on small farms than on large farms, it represents a significant proportion of the total received even on high-income farms. Members of farm families have to compete in the same labour markets for the off-farm proportion of their incomes as the rest of the population. In order to compete they must have at least the same level of skill and education.

In rural areas, there is need for wealth - generating activities other than agriculture to utilize fully the labour available, particularly on grain farms, and to provide greater economic stability to these areas. Progress is being made as service and other industries expand in these areas. One such example is the location of production of newtechnology farm equipment in rural areas that possess the necessary infrastructure.

The above data indicate that on average about half or more of the income received by farm operators comes from nonfarm sources. The proportion declines as income rises. The proportion of operator income received from off-farm sources remains significant, even for those operators having incomes of $250,000 or more where the off-farm income received exceeded $37,500 per operator. The proportion of off-farm income declined as farm-operator income increased, the range for the individual operator income classes being 58.0 to 24.2 percent in 1990 and 59.7 to 38.0 percent in 1991.

Rural communities have been under considerable stress in recent years. Many have been in a decline evident for decades. There are numerous reasons for this decline, some at arms length from the price of grain. A primary reason for the decline has been the substitution of capital for labour on the farm. Mechanization has increased productivity per person enabling consolidation of farm units with an attendant out-migration of people formerly on the land. The recent change in grain production technology is the latest example of a long line of similar developments. All have resulted in less people remaining on the land.

Fewer people left on the land reduces the potential customers for businesses in the local towns and hamlets. However, another factor has also been at work. That is the advent of continuous improvement in the road transportation system. Coincident with this improvement has been a change in the shopping patterns of rural dwellers who have developed urban tastes which can only be satisfied at larger centres where a greater variety of goods is available at competitive prices. Rural dwellers are now prepared to go to the larger centres to obtain the services desired and in the process bypass the local centres, eventually leading to the demise of the latter. Thus while these local centres have suffered the major centres have expanded.

Perhaps in no other area is consolidation more evident than in the school system. The demand for provision of improved school services at an acceptable cost has given rise to the busing of students from rural areas to central towns or cities. The role of the school as the centre of many farm communities has now disappeared. As well, many rural post offices have closed.

Provision of infrastructure in rural areas initially absorbed much of the labour surplus to agriculture. Other local industries such as farm machinery manufacturing also provided jobs along with other activities. However, over time jobs from these sources tapered off as service industry jobs increased. This did little to stabilize rural communities.

Recently, however, two developments changed the dismal prospects of some local communities. The desire of persons in the city for a more tranquil living environment has resulted in communities adjacent to the cities becoming "bedroom communities". In addition, telecommunications have taken over service tasks in remote areas formerly performed by people. Examples include insurance record keeping and payroll services. Improvements in computer controlled automation have enabled specialized farm production units, such as those for broiler production, to achieve the economies previously only attainable by large "factory in the field" types of operations.

The depopulation of rural areas has major implications on local services for those remaining. The previous local social centres, such as the school, the church and the community hall have been replaced by those more remote. Health services have likewise changed, with doctors and hospitals being located in the larger rural centres. The services remaining tend to be at a distance and entail additional transportation costs.

The decline of the local community has been said to bear a relationship to the sustainability of agriculture. Such a relationship can only be considered indirect as it reflects changes in farm technology and markets which may or may not be conducive to sustainability. The new rural map being drawn as a result of the redistribution of population on the plains reflects changing tastes and social structures. In as much as the redistribution is the result of economic forces, it can be stemmed only by greater employment opportunities in rural areas. While nostalgia may prevail for rural living, government policies should be built on economic opportunities which foster diversification in these areas. The situation in Saskatchewan provides valuable insights into the impact of greater diversification in agriculture upon service centres. Southern Saskatchewan is marked by either large farms following a grain monoculture or large scale ranches. Both require relatively few individuals to operate with the effect that population declines.

A somewhat different situation prevails in northern Saskatchewan. Agriculture there is more diversified, with the population increasing. This growth becomes important both for the well-being of service centres and for the development of manufacturing. However, evidence of any reversal of a decline in trade-centre status as a result of manufacturing is uncommon. The same situation applies to mining operations at a distance from major wholesale/retail centres. Manufacturing or mining centres in relative proximity to the major wholesale/retail centres take on the attributes of bedroom communities without a commercial structure.

The same governments selectively attempted to offset the negative effects that inevitably followed by retaining unused or seldom used facilities or distributing infrastructure investment so that expenditures were made in as many communities as possible. In the process, consolidation was accelerated while funds were wasted which could have been used to ensure a complete complement of infrastructure at selected locations. While it would be theoretically possible to create employment opportunities in enough rural communities to slow or prevent further consolidation, this option is not realistic.

Pursuit of cost-minimizing efficiency in infrastructure investment would be less expensive than the course followed previously but would result in further consolidation of rural communities. By coordinating the efforts of governments and private organizations, the preservation of more rural communities without a major sacrifice in efficiency would become possible. Stabler et al are partial to trying the coordination approach in an attempt to preserve more rural communities at a minimum cost to society.

During the 1982 to 1991 period, the value of agricultural exports represented from 36.5 to 48.9 percent of farm cash receipts. Most of the agricultural exports emanate from the prairie region. Income from exports of grain in particular has suffered in recent years from the export subsidy practices engaged in by the EEC and the U.S. While agriculture has been dealt with in the present (Uruguay) Round of the GATT negotiations, interest remains high among producers. Some of them foresee improvement in their competitive position. Others, primarily those operating under supply management who were concerned that their privileged position might be eroded, have been partially satisfied at least in the short run by the protection of tariffs at levels which others consider "obscene".

Success of the Uruguay Round of multilateral negotiations hinged on a successful agreement being reached for agriculture. In order to expedite progress, Arthur Dunkel, when Director General of the GATT, took a direct role in the negotiating process. His proposal became the basis for the subsequent negotiations on agriculture. Rude, Massow and Martin have provided an invaluable summary of his proposal and an assessment of its effects if adopted. The GATT is based on the philosophy that market forces should be the primary factor driving international economic adjustments. The basic objective is the removal of impediments to trade.

The GATT is a commercial agreement consisting of 38 articles which spell out the guidelines under which trade is to be conducted. The first part of the agreement is built on two principles, namely, that signatories not discriminate against the products of another member (most favoured nation (MFN) treatment) and that members do not raise import tariffs above agreed levels. Import quotas are condemned in the GATT charter. Attention focuses on eliminating import quotas and other non-tariff barriers and replacing them with tariffs.

The second part of the agreement impacted heavily on agriculture since it dealt with prohibition of quantitative import restrictions, customs regulation, state trading, antidumping and countervailing duties, dispute settlement and escape provisions. Two articles of this agreement were of particular importance to Canadian agriculture: Article XI which prohibited import quotas except where domestic output of agriculture and fishery products was restricted; and Article XVI which prohibited export subsidies except in the case of agricultural products. Protection for Canada's supply management programs was provided under Article XI©(1), this being initially introduced to allow the U.S. to protect its farm programs under Section 22 of the U.S. Agricultural Adjustment Act. The EEC was allowed to use variable import levies, these not being specifically excluded under the GATT. Furthermore, the GATT defined no limits on domestic assistance to agriculture.

These exemptions allowed developed nations to provide high levels of support to agriculture. The result was that surplus grain stocks accumulated, international prices were depressed, trade disputes developed, and trade patterns were created which were inconsistent with market forces. Furthermore, new areas were negotiated, these including services, intellectual property and textiles. The negotiations therefore became very complicated. Some countries vowed that if progress was not made on agricultural trade they would not sign the document. Dunkel attempted to overcome the impasse which developed in negotiations by putting forth his proposal. Its provisions extended to four broad categories: market access, domestic support, export competition, and sanitary and phytosanitary regulations.

1. *Market Access*: There were four components to market access.
    1. Conversion of all non-tariff border measures to tariffs (tariffication) equivalent to the level of import protection during the 1986 to 1988 period.
    2. Average tariff reductions of 36 percent across all agricultural products during the 1993-1999 period with the minimum reduction of an individual tariff of 15 percent.
    3. Minimum market access to imports using a tariff rate quota mechanism allowing a certain proportion of a country's domestic consumption being available to imports at a preferential tariff rate; equal to 3 percent of domestic consumption initially rising to 5 percent by 1999.
4. A safeguard mechanism to temporarily limit import surges or to take into account price declines.

2. Domestic Support Provisions

Domestic programs were classified as "green" or "amber". "Green" programs were permissible and not subject to mandatory reduction. "Amber" programs were inconsistent with GATT principles and expenditures on them would have to be reduced. "Green" programs were those which were generally available to all forms of agricultural production. "Amber" programs were lumped together and their financial cost measured in terms of an aggregate measure of support (AMS). "Green" programs provide price support to producers other than those in such areas as research and extension, disease control, food aid and direct payments to producers not related to either price or output. "Green" programs are considered non-countervailable. Measurement of the support of "amber" programs was defined in terms of price support, direct payments and other payments such as input subsidies and measures to reduce marketing costs. Expenditures on "amber" programs must be reduced by 20 percent from 1993 to 1999, a maximum limit of assistance being established.

3. *Export Subsidies*

Export subsidies must be reduced on agricultural products. This reduction will bring agriculture under a discipline similar to that for other products. A prime example of such a subsidy is that provided under the Western Grain Transportation Act. During the six year period, 1993 to 1999, expenditures on export subsidies are to be reduced by 36 percent from the 1986 to 1990 period, and the volume of product on which export subsidies are paid must be reduced by 20 percent.

4. *Sanitary and Phytosanitary Provisions*

Under Dunkel's proposal, an attempt was made to limit use of health and sanitary regulations as non-tariff barriers. The thrust is toward working toward international harmonization of standards and practices. The proposal also contained a dispute-settlement mechanism, The Understanding on Rules and Procedures Governing the Settlement of Disputes, which would be of major benefit to agriculture. Furthermore, adoption of the Dunkel proposal would shift agricultural trade policy away from a protectionist course.

Most importantly, it is time the uncertainty ended so that Canadians can get down to the business of demonstrating their ability to compete." Agreement on Dunkel's proposal would have required existing national agricultural policies to be compromised, including those of Canada. The EEC and the U.S. could not agree on adopting Dunkel's proposal as written. They did agree to overcome some of their differences in the agricultural component of the Uruguay Round and to reach a settlement on their oilseed dispute. The agreement was termed the Blair House Accord (BHA). In the process, changes in the final draft of Dunkel's proposal were agreed upon. These changes related to domestic and export subsidies.

Analytical Methodology

The meaning of sustainability remains a subject for debate, particularly in the field of agriculture. Its many dimensions include economic, social and environmental matters. An integrated approach is therefore required if true sustainability is to be achieved. Definition of sustainability becomes difficult since philosophical and value judgments are involved. The value accruing to society from sustainable development is difficult to measure, particularly for social issues such as the quality of life, equity and poverty alleviation. Indeed, these issues involve value judgments which tend to be influenced by theological views. The question becomes whether the correct end and the means for achieving that end can be determined by using the traditional tools for measurement. Traditional analytical tools nonetheless continue to have their place in measurement of the many other aspects of sustainability.

Even in the social realm these tools have much to contribute. Measurement of sustainability in agriculture is rendered more difficult by the dichotomy of views held by ecologists and economists. Going in one direction are those who analyze sustainable agriculture within the framework of ecology. Going in another direction are those who analyze sustainability within the framework of traditional economics. There may be said to be five major differences between these two disciplines: the time frame normally considered; the degree to which nature as well as humans enter into the analysis; the attitude towards natural and human created capital; the degree to which non-market goods and services are incorporated into the analysis; and the degree of emphasis placed on environmental quality. These differences in views are not insurmountable since both of these disciplines are interested in increasing sustainability.

There is therefore value in using available tools for the measurement of sustainability. Pearce and Warford have used these to good advantage in their analyses. There is a substantial body of knowledge with respect to measurement of the various attributes of sustainability including welfare, public choice, market failure , externalities, private and public goods, and intergenerational equity. Contributions have been made by a large number of researchers to the measurement of sustainability. A distillation of the individual contributions is therefore presented below. These contributions extend to most aspects of sustainability.

Welfare is an essential aspect of sustainability. It is of prime importance that the welfare of the community as a whole be maximized. Welfare is linked to development. Increased output gives rise to a greater availability of goods to all members of society thereby increasing aggregate wealth. Increased output may in turn impact on the environment. Furthermore, the distribution of goods and services among persons gives rise to questions of equity. Development therefore becomes highly relevant to sustainability. Many contributions have been made to the measurement of welfare, only a few of which will be mentioned here. Reder gives substance to the meaning of welfare in his indicator that welfare increases (decreases) whenever one or more individuals become more (less) satisfied without any other individuals becoming less (more) satisfied. He also introduces the concept of compensation being paid to the individual being adversely affected by the change, this compensation being made available by an equivalent tax on the individual benefiting - total welfare increasing if the net benefit of the change exceeds zero. Arrow argues that welfare increases even in the case where compensation is possible but remains unpaid. Little further expands Reder's concept and states:

> "A change is economically desirable if it results in a good distribution of welfare, and if a policy of distributing money by lump sum transfers could not make everyone as well off as they would be if the change were made."

When production or consumption activities give rise to side effects which may be of benefit to others or inflict costs upon them, these effects are called externalities. Their distinguishing feature is that they are economic creations which do not or cannot go through the market. As a result, some persons receive benefits at no cost to themselves while others incur costs or endure discomforts without compensation.

From an economic point of view, the issue raised relates to efficiency. Decisions should be made after assessing all the costs against all the benefits, regardless of whether these costs are determined by the market. The costs and benefits of externalities are seldom if at all determined in the market place. To achieve true efficiency such costs and benefits (externalities) must be taken into account (internalized). Activities which produce beneficial or positive externalities, in order to ensure an adequate supply of the service is available, are often provided by public intervention. The situation is much different for activities which give rise to negative externalities. Internalization of such externalities implies either taxing the gainers or compensating the losers. These externalities may be of the nuisance category such as noise or even a dust storm caused by improper farm practices; of a capacity category where use of a resource by one person may adversely affect use of the resource by others, as in land used for agriculture versus wildlife habitat; or of a supply category where only a finite resource is available, thereby affecting use by future generations, as in the case of mining. The impact of the latter on future generations m ay be mitigated by advances in technology which effectively increase the supply of the resource or provide substitutes.

Scitovsky considers negative externalities in terms of the impact of production practices on the environment. He draws attention to issues which have come to be better recognized in the present day. These are the social costs, wastes and pollution associated with production yet not internalized. He introduces the concept of pollutants reaching a critical level beyond which the regenerative powers of nature are unable to cope, a concept so eloquently described by Holling. Environmental externalities are said to be the most difficult to internalize with the state often having to intervene on behalf of the public since the individual responsible for the pollution or other externality may not be readily identified.

Nathdevotes considerable attention to the social objectives of society while recognizing the existence of externalities. These objectives vary over time. Achievement of these objectives is attained through policy instruments including taxes, subsidies, direct public investment, the pricing policies of public enterprises and, to a degree, through the actions of private organizations. An individual policy instrument may affect more than one objective, and an objective may be affected by more than one instrument. He expresses economic efficiency in relation to what he describes as social welfare functions. He contends that attaining technological efficiency and being continually innovative are desirable goals, as they would help obviate many resource allocation problems.

Though it isn't expressly stated in his work, Nath is a strong supporter of research and development. Rowley and Peacock devote considerable attention to market failure, the failure of the market to take into account all costs including those associated with externalities. They categorize goods as being of three kinds: private goods where consumption by an individual deprives others of that use, these goods being characterized by production cost as well as consumption cost; public goods which are enjoyed by all, an individual's consumption having no effect on that of others with no cost attached; and mixed goods which possess the characteristics of both public and private goods.

Externalities are associated with mixed goods and these give rise to market failure. The two authors introduce public choice as a decision-making tool when dealing with externalities. They devote considerable attention to the problem of pollution and propose a tax rather than a regulatory scheme to maintain water quality. The procedure offering the greatest promise is the establishment of a value for each unit of pollution production, with the water quality entity issuing a share equal to this value. These shares would be offered at public auction, the total number of shares issued reflecting an agreed upon acceptable level of pollution.

Polluters would be required to hold shares equal to their level of pollution, the cost of these shares providing an incentive to adopt available technology to eliminate such pollution. Turner expands upon some of the earlier ideas and divides capital into two parts, environmental and human created. Both provide a stream of goods and services. The natural environment provides among other things:

1. A source of raw materials for production processes.
2. A sink for the waste products of production processes.
3. Life supporting services.
4. Aesthetic goods.

Environmental capital may or may not be substituted by human created capital. It may be depreciated, augmented or liquidated and be used directly or as an input to production processes. Some think natural resources tend to befinite.

Technological improvements, discovery of substitutes and development of new reserves stimulated by rising prices of conventional sources have often resulted in an increase in the abundance of resources, the replacement of coal by natural gas being but one example. These events lead others to believe that the resources available become limitless albeit at increases in cost. Others argue that finiteness has already been reached. Logic would suggest that the true situation lies somewhere between these two rather extreme positions.

Measurement of the level of natural or environmental capital is required if the true costs of production processes are to be determined. Attempts to determine this level have led to pressure to expand National Income Accounting to include valuation of environmental goods and services.

Such a valuation is difficult since these are non-market goods and therefore do not have explicit prices. Several approaches have been taken to achieve an evaluation, including: replacement value; property value; contingent valuation; and total economic value. Non-pricing of environmental goods leads to their ove ruse. But the difficulty of costing environmental goods and the politics of making any price effective frustrates efforts to incorporate their cost into a production process.

The inability to value environmental goods with any predetermined degree of accuracy renders a true economic benefit cost analysis difficult to achieve. The issue becomes one of determining the safe minimum standard for use of the environment in order to take into account sustainability. This idea was introduced by

Economy is an ingredient of sustainability along with environment and the well being of people. Four factors present themselves in this regard:

1. Environmental damage is of importance not only because of its economic effects but because it lowers productivity, soil degradation being one example.

2. A greater priority must be given to the environment if economic policies are to be sustainable, such as in agricultural policy.

3. Attention is required to ensure that economic growth is also sustainable growth, such as in the emphasis placed on the production of wheat on the prairies.

4. While raising income continues to be a major goal of policy, the need remains to be aware of the potential trade-offs between increases in income and environmental deterioration, with poverty understood to be a major cause of environmental destruction.

Costs are of paramount importance in production. The relationship of costs to output is reflected in the volume of output which will be made available at alternative prices.

Costs may be expressed in terms of the values of the items foregone as a result of using particular inputs in production of a given product. Costs may also be expressed in terms of the funds required to produce the desired product. These costs include those which involve an actual financial outlay and those which do not. The latter costs may be considered in terms of their return if used elsewhere. Time is also a factor in production as it allows plant size to be optimally adjusted in terms of costs and output.

Sustainability requires that all costs be recognized in production. Such costs include the impact on the environment, the effect of present use upon future availability, use of common property, unpriced inputs and even non-use values. Unfortunately, accurate determination of many of these costs is very difficult if not impossible. It is nonetheless essential that these externalities be costed in order to determine the true cost of production of any commodity.

For agriculture to be sustainable all the costs associated with production must be identified and determined whether in the form of direct outlays or of externalities. Establishment of these costs will enable a direct assessment to be made of the economic viability of a particular form of production given the return available. Sustainability may be defined in terms of the flow of income from production. Issues to be considered when measuring sustainability include the discount rate; private versus social costs (detrimental effects or negative externalities); non-market benefits; economic flexibility and income/risk preferences.

At the same time a producer surplus is required to sustain farm families. Means to assess each of these relative to agricultural production are discussed below.

1.Discount Rate

Equivalent opportunities should be available to future generations as experienced by the present generation. As applied to agriculture, this requirement infers that equivalent resources be available for use in the future. Equity may be considered to prevail where the prices of inputs used at present are equivalent to those which would be experienced in the future, given a particular discount rate. The price of the input in the future cannot be accurately foretold, since technology may effectively increase the supply available.

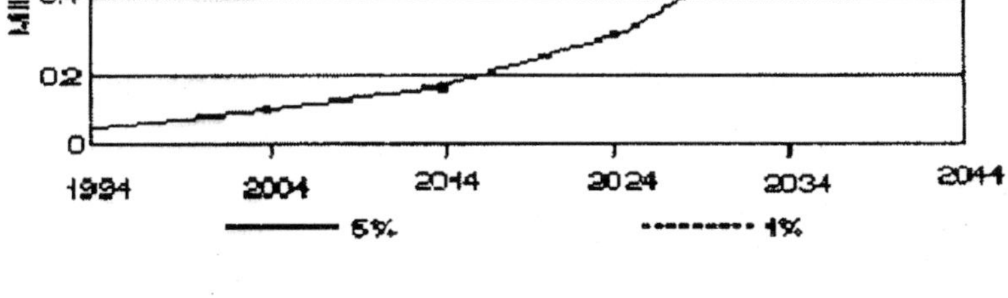

Figure 3. Discounted Value at Interest Rates of 1 and 6 Percent of $1 Million in 2044 at Ten Year Intervals, 1994-2044

On the other hand, a particular resource may be capable of being finite, for example wetland contiguous to agriculture which is expected to increase in value over time.

Establishing an appropriate discount rate is of significant importance. An input in limited supply is considered to have a high future value. Therefore, in order to encourage restricted use in the present, a low discount rate should be imposed, resulting in higher prices in the market place. The use of energy in agricultural production offers an ideal example. Rather than subsidizing the cost of agricultural fuel, it would appear more appropriate if the fuel was costed in terms of its perceived future value in order to minimize present use and encourage the development of new sources of energy.

An illustration of the impact of alternative rates of interest upon the discounted value of a resource is provided in Figure 3. If the resource is valued at one million dollars 50 years hence, its present value, assuming a six percent interest rate in the interim, is $54,290. If it is assumed that a one percent interest rate prevails in the interim, the present value is $608,040. The importance of the discount rate used, once the future value of the resource is established for its valuation at present and intervening years, becomes very apparent.

2. Negative Externalities

Effects considered detrimental associated with agricultural production include:

1. Herbicide, pesticide and fertilizer residues
2. Degradation of the soil resource, loss in organic matter
3. Destruction of habitat, particularly that of wetlands
4. Excess use of water
5. Contamination of air and water

6. Output of greenhouse gases associated with use of fossil fuels

7. Interference with the natural ecosystem

Methods of Costing

Producers use herbicides, pesticides and fertilizers as aids to production. While the costs of these items are established in the market place, the detrimental effects arising from their use are not costed. Judicious use will minimize any detrimental effects. Furthermore, many of these items are benign to the environment. As an option for those which do have detrimental effects, their true cost could be established by ascertaining the crop losses which would arise from non-use and deducting the original cost of the chemical. This sum could be offered to the farmer in exchange for not using the chemical.

The cost of production in terms of degradation of the soil resource is difficult to establish but easy to identify. One of the proxies for soil degradation is the loss of organic matter. While inept farm practices in the past have given rise to soil degradation, means are at hand for ensuring the present level of organic matter in the soil is maintained or enhanced through cultural practices which provide continuous trash cover. The decline in soil productivity under cultivation has been overcome by the use of technology. Presently, the replacement cost of the soil nutrients as fertilizer may be considered representative of the impact of production on the soil, with this cost included in farm budgets.

Agriculture has a major impact on the habitat of birds and animals. As a result, some species thrive while others decline. In recent years, government policies in combination with the "development of the 40-foot cultivator" have had the effect of encouraging the draining of sloughs and potholes on the prairies and thereby affecting waterfowl habitat. Earlier maintenance of this habitat by the farmer was a positive externality enjoyed by the public. The cost of maintaining this habitat is represented by the amount of money which would have to be paid to the farmer to maintain the wetland and for any crop damage associated with the greater bird population.

Agriculture is by far the greatest user of water on the prairies. Approximately 88 percent of this water is used for irrigation. Unless this water is wisely used, it can contribute to soil salinity. The low prices charged for water encourage excessive use. While most water development projects on the prairies have benefit/cost ratios in excess of unity, there is a tendency not to charge market clearing prices for water in order to encourage irrigation development. However, as pressure increases on the available water supply there is a need to price this resource according to what the farmer is willing to pay for the water as reflected in the net benefits received from irrigation development.

Confined livestock operations can result in contamination of the air and the water supply. Most of these operations are now required to be benign to the environment through regulation. In these cases, the costs of conforming to the regulations are included in the total costs of production. On the other hand, the effluent and manure produced if properly handled can be a valuable joint product of the operation by increasing soil tilth and fertility. In addition, the technology is available to capture the methane produced for use as fuel and in power generation. The value of these joint products may exceed the costs of any environmental regulation required to make these operations neutral in terms of air and water contamination.

Contrary to the perceptions of many on the prairies, agricultural production uses only about 4 percent of Canada's total energy consumption. Direct energy use includes gasoline, diesel fuel, electricity and heating fuel. Indirect energy inputs are those used in the production of fertilizers, seed, pesticides, herbicides and the energy component of machinery. There is a major potential for reducing the energy used by adopting less intensive tillage systems. In theory, agriculture has the potential to produce more energy than it utilizes. While the cost of energy is significant in agricultural production, the issue of whether the price in the market place relates to real cost remains. Price can be used to encourage conservation and also to lead to an increase in supply. Energy taxes, by increasing price, have this effect. On the other hand, eliminating taxes or even providing subsidies on farm use of fuel have been shown to have the opposite effect. The actual cost of energy is incorporated into the cost of agricultural production. Whether the true cost is thereby reflected remains a moot point.

3.Positive Externalities

The positive externalities with respect to agriculture remain largely unpriced. These include such things as value to others of the rural landscape, including the presence of wildlife, and an attractive lifestyle. Most of these benefits involve interpersonal evaluations and cannot be said to reduce agricultural production costs. It is possible to utilize a surrogate market technique to establish the net value of country living. Where housing of the same quality is available both in a city and a rural setting, the net additional costs associated with country living - increased travel, a different tax base, local recreational and other services, and less free time - become a measure of the value of a rural lifestyle.

4. Risk and Uncertainty

These add to farm costs and where commercial risk-taking or insurance firms establish premiums, the premium becomes an input cost. Not all risks and uncertainties can be offset by insurance. Agricultural production is subject to the vagaries of nature. Receipts from the market place vary greatly over time. The result has been agricultural policies which attempt to stabilize returns and to offset losses arising from natural causes. None of the policies developed thus far have been entirely successful. However, the reduction in risk exposure has encouraged an increase in the size of agricultural production units. These units tend to have lower costs of production when appropriate adjustments are made in their land, labour and machinery components.

5. Income

While the discussion thus far has dealt with the cost of agricultural inputs in a sustainable environment, it is essential that the returns from production exceed the costs in order to sustain the human family on the land. Producers receive little if any income from the positive externalities they provide to society. Consequently, they have to depend upon the prices available in the market place. Since grain producers on the prairies are so dependent on the export market for their grains, the predatory pricing tactics of competing exporters have reduced their net incomes below that necessary for a livelihood. When assessed in this respect, grain production is presently unsustainable. This condition cannot continue because when the very basic necessities of life are placed in competition with the environment, the environment will always lose. The situation may be alleviated somewhat by the new GATT agreement and also by a restructuring of grain agriculture on the prairies. The situation is different for producers of livestock and several other farm commodities where the net returns from production are sustainable.

For sustainability, both the costs of inputs established in the market place and also those for which no market exists should be included when costing production of agricultural products. The market cost of inputs is easily determined. The same is not true for inputs for which no market exists. Various devices have been illustrated as a means to estimate these prices. In some cases, no explicit value can be expressed since such a value depends on the viewpoint of the individual. The income obtained from sale of products must, however, exceed their production cost by a significant margin in order to provide the farm family with an acceptable income.

Sustainable agriculture is a subset of sustainable development. Sustainable development can be defined as:

> "... a process of change in which the exploitation of resources, the direction of investments, the orientation of technological development, and institutional change are all in harmony and endorse both the current and future potential to meet human needs and operations."

This definition has come to be accepted in many areas of sustainable development research. It provides an idea of the scope of sustainable development, but does little to explain how sustainable development can be operationalized. Sustainable development can best be described as a concept rather than being strictly defined as a practice. This concept must be implemented at all levels and areas for sustainable development to be realized.

2

Land Management

Land is the solid part of the earth's surface. Its value in agriculture is tied to its capacity for providing services used to produce an agricultural product. Part of land's agricultural use value is a result of the climate that dictates the length of the growing season, available rainfall, and other determinants of the crops that can be produced and the risk involved.

Land has at least two distinguishing characteristics. The first is its location fixity. As a result of land's location fixity, if the control of land is to be transferred, the users of the land must relocate their production activities. Land's second distinguishing characteristic is its ability to maintain over time and use its physical qualities that give it value. Land that can exist for an infinitely long time without significant deterioration is referred to as an infinitely lived durable.

Most analysts recognize that the two properties of land, location fixity and long if not infinite life, are approximations of reality. For example, land in sod farms is indeed transferrable. Moreover, lack of proper maintenance may result in land losing some of its physical properties that give it value in a short period of time. Still, land serves as a reference point for defining an asset whose physical properties are long lasting and cannot be moved except with considerable effort and cost that is usually considered prohibitive.

Land is differentiated from real estate inasmuch as real estate includes capital structures attached to the land. Capital structures such as buildings, grain bins, and fences are durable assets that can be removed and their services obtained at a different location. These types of structures are treated as durable assets and are not included in the costs or services of land.

Other capital structures that are a part of real estate such as wells, terraces, levees, leveling, and tiling become part of the land and acquire its properties of being immovable and durable. These investments are not transferrable but depreciate over time. Because services of these investments cannot be transferred to another location, even though they can be removed (negative salvage value), they are usually treated as part of the land. When rental rates are used to estimate the cost of land use, the cost of these investments will typically be included in the rental rate.

A distinction between land's agricultural use value and its market value is required to measure costs and returns (CARs) associated with agricultural land use. It is also important to recognize that in some cases, the agricultural use value of land differs from its market value. One explanation for the difference may be the expectation that land will be used in nonagricultural activities beginning at some future date. For example, land close to an urban center may have a high price based on expected value in residential housing.

Another explanation is that land earns nonagricultural returns concurrently with agricultural returns as may be the case where a producer is able to extract a payment during winter months for hunting or trapping. The importance of distinguishing between land's agricultural use value and its market value is that not all CARs associated with land should be charged to land's agricultural use.

Ownership of land entitles an individual to the control of the sum total of natural and manmade resources associated with the land. What is being controlled is a resource with value that if liquidated would, with the passage of time, generate a return. The value of this resource, as has been stated already, is its current agricultural use value. The foregone return associated with land's agricultural use represents land's opportunity cost. Conceptually, the opportunity cost of land would be approximated by multiplying the current agricultural value of the land by an appropriate interest rate, thus giving a perpetuity with the same present value as ownership of the land's agricultural use value.

Finally, opportunity cost is not the same as the cost of financing land. Financing land purchases is not part of land ownership costs because land purchases reflect alternative means of acquiring or controlling land and are often tied to acquisition costs rather than use values. Moreover, trade-offs are often made between the sale price of land and the interest rate charged, as in some land contract sales. Besides opportunity cost, property taxes are the other major cost of owning land. Consistent with the distinction between agricultural use value and land's market value, only the portion of property taxes reflecting land's agricultural use value should be included as an agricultural land cost. Several states have no property tax.

Many other states already have some form of a greenbelt tax law in which agricultural land is taxed at its agricultural use value. The maintenance costs of holding a capital asset such as land are the expenses required to maintain the service potential of the asset at a reasonable level and to extract services for a single time period. Maintenance costs for land are thus those costs required to restore land to its approximate service potential at the beginning of the period or to prevent its deterioration during the period.

Whether or not they are actually incurred is immaterial. They represent costs associated with actual deterioration or the cost of preventing deterioration of land and should be accounted for in cost estimates of using and controlling land. Land incurs service reduction costs because the service capacity of land is altered as a result of use or the passage of time. Costs required to restore service capacity as a result of use are referred to as user costs. Losses in service capacity as a result of the passage of time are referred to as time costs.

Estimating Land's Agricultural Contributions

In practice, it may be difficult to estimate separately the opportunity costs, user costs, time costs, property taxes associated with agricultural use value, and other overhead costs of holding land. Property taxes associated with the agricultural use value of land require precise measurement of land's agricultural use value, which is not easily obtained. For the same reason, we are often unable to estimate precisely opportunity costs associated with land. Finally, use or time costs are difficult to measure when land markets are thin and do not provide a sufficient number of observations to make reliable estimates. At the same time, there are active land markets in many areas, and in areas with less active markets, rural appraisers, farm management consultants, and local producers may be able to approximate the agricultural value of land currently being used for agricultural production.

A problem similar to separating land's agricultural use value from its market value is encountered when decision makers desire to construct enterprise CAR estimates when more than one enterprise share the use of the land. Because land costs are frequently calculated as a residual measure, any attempt to allocate returns from enterprises using the same land allocated between the several factors of production including labor, management, risk, capital, and land will be arbitrary.

Land Tenure Considerations

Access to rights to use of land may be gained through alternative tenure arrangements. These may be acquired through ownership, cash rental for a specified period, or share rental for a specified period and use. The tenure method used to acquire rights to use of land affects production risks, production costs, and production returns to both the entrepreneur and land owner. Modal tenure patterns for rights to land use differ greatly among geographic areas and among land quality within areas as well as among production enterprises within areas.

The owner-operator receives the full return from production of the enterprise, experiences all cash costs associated with production, bears all production risks, and experiences both cash and noncash costs of land ownership and maintenance. The owner-operator views the costs of land as a fixed residual cost, and treats them as such in production decisions. These land costs include all land directly associated with production enterprises as well as land associated with the firm as an entity used for service areas and waste lands not suitable for production.

The cash rent tenant-operator receives the full return from production of the enterprise, experiences all cash costs associated with production, including a cash cost for land use, and bears all production risks. The land use cost is thus a variable cash cost to the cash rent tenant-operator and is treated as such in production decisions. The cash rental charge may reflect only production acreage or total land area in the farm unit, depending on whether the contract was for production of a specific enterprise or for the farm unit including service areas. It is extremely important to determine whether a cash rental rate is for tillable (and expected to be planted) acres or for total acres including acres not devoted to production because this will affect the cost of land per productive acre.

The landowner renting the use of land on a cash rent basis experiences no production risks and receives a fixed rate of return for the time period specified in the rental contract, subject to variations in supplemental returns associated with program payments for government-supported crops. The costs of land to the landowner who cash rents production rights to a tenant are reflected as a fixed residual cost to the landowner, subject to variations in maintenance costs that may be specified in the cash rental contract.

Share rental tenure arrangements are much more complex and vary greatly among production enterprises, land quality, and geographic areas. Share rental arrangements substantially alter both cost structure and production risk for both landowner and tenant. Both landowner and tenant share production returns and production risks, the proportionate shares for each varying by production enterprise, land quality, and proportion of cash production expenses shared. Production risk may be altered by the proportionate shares of selected cash production costs borne by each.

Thus the share tenant-operator receives only part of the production returns and experiences land use costs as a variable reduction in receipts. The tenant-operator bears all or part of the cash production expenses, dependent upon terms of the share rental contract, and faces a substantially altered marginal cost-marginal returns structure compared to the owner-operator or cash tenant-operator. The landowner in a share rental tenure arrangement is a much more active participant in agriculture production activities than the landowner in a cash rental tenure arrangement. The costs of land use are reflected to the landowner in a share rental tenure arrangement as a fixed residual cost, and are treated as such in production decisions.

Estimating Land Costs

All the land cost estimation complexities described in the previous section lead us to look for alternative ways to estimate costs associated with the use of land in agricultural production. Since land tenure patterns differ substantially among regions and type-of-farming situations, it may be appropriate to use different procedures for estimating land costs in order to minimize bias in land cost estimates among situations.

Cash rent paid for land used to produce an agricultural product best measures the sum of opportunity costs, time costs, user costs, property taxes, and other overhead costs associated with land's agricultural use value in those situations where a significant proportion of the agricultural land is farmed under cash rental tenure. Cash rent paid for the use of land in agricultural production is the amount of compensation the land owner requires from the tenant to pay for property taxes on the agricultural use value of the land, opportunity costs, time costs, and user costs. Because the tenant does not receive capital gains on the land, cash rent does not include the value of anticipated gains (losses) due to inflation or potential future nonagricultural use of the land.

Rental rates also reflect what tenants are willing to pay to avoid the payment of property taxes on the land they intend to use for producing agricultural products and to avoid the payment of opportunity costs (on the agricultural value), time costs, and user costs. Cash rent does not include payments to financial capital, risk, and management, because in most cases the tenant is not acquiring them and would not be willing to pay for them. Only if the landlord absorbs some of the production risk, supplies capital, or provides management does the rental rate include charges for these factors of production.

In a share rental agreement, for example, the landlord is sharing in the production risk and the payment he/she receives must compensate for the risk. But in a typical cash rental agreement, the tenant is acquiring only the temporary use of the land to produce an agricultural product and therefore is willing to pay for only the land's services used in agricultural production during the production period. Thus, the rental rate willingly paid by the tenant and accepted by the landlord is the best estimate of land's agricultural use for use in CAR estimation. Cash rental rates are probably the most reflective indicator of current market conditions in agricultural production areas where a significant proportion of the agricultural land is farmed under cash rental tenure. However, these estimates suffer bias resulting from a number of factors. Cash rental rates usually apply to the use of the land for a specific agricultural production enterprise, with the rental charge differing among enterprises. All agricultural production enterprises do not share proportionately in the land rental market, whether it be cash or share rent.

Some enterprises are predominantly cash rent whereas others are predominantly share rent, differing by geographic area. In some geographic areas the cash rental market represents such a small portion of the agricultural land that cash rental rates do not reflect production values of land, but rather tend to reflect land of marginal quality or, frequently, favorable treatment in family land use transactions. As mentioned previously, it is important to determine what is included in "one acre" of land. The effective cost to the production enterprise will differ depending on the whether this acre contains grass waterways, extensive headlands, irrigation ditches, buffer strips, or windbreaks, power lines (or substations), and/or barren hilltops.

In some situations, share rentals may provide the most accurate reflection of current market use value for land, particularly in areas where share rental is prevalent for particular production enterprises. Because share rental arrangements are quite flexible, differing among individuals in the proportions and specific production costs shared, modal or typical rental rates are more difficult to identify and quantify. Estimates will suffer from bias because, frequently, the cost sharing consists of cash costs for the landowner and both cash and noncash costs for the tenant. While share rental rates can be converted to a cash-equivalent value, they should not be treated as cash rents because they affect CARs for both the tenant and the landlord.

The following example illustrates the calculation of a cash-equivalent value for a share rental arrangement. Assume a two-fifths crop-share agreement for soybean production where the landlord receives 40% of the receipts and pays 40% of the cost of seed, pesticide, and insurance. Further assume a soybean yield of 42 bushels per acre valued at $5.50 per bushel, and seed, pesticide, and crop insurance expenditures amounting to $27.08 per acre. The calculated cash-equivalent land rental value would then be $81.57 [(0.4)(42)($5.50) - (0.4)($27.08) = $81.57]. In this situation the share tenant experiences a reduction in cash receipts of $92.40 and a reduction of $10.84 in cash operating costs in lieu of a cash rental payment.

In areas where farmland is operated almost exclusively by owner-operators, rental rates may not provide meaningful estimates of fair market value. When reasonable estimates of the market price of land for agricultural purposes can be obtained, an implicit annual rental fee can be obtained using standard capitalization techniques. The market price of land for agricultural purposes times the real interest rate (adjusted for risk), plus annual maintenance costs, plus annual net real estate tax will provide the least biased estimate of annual real use cost.

A real interest rate is appropriate for capitalization because the Task Force suggests that all CAR estimates be in real (and nominal) terms as of the end of the production period, and that all CAR streams outside of the current period be estimated in real terms. If the agricultural land value is in beginningof-year terms, it should be multiplied by the current inflation rate to put it in end-of-year terms before computing the opportunity cost equivalent. As an example, consider an acre of land with an agricultural value of $2,163 per acre as of the end of the year. If the real interest rate adjusted for risk in agriculture is 5%, then the implied opportunity cost is $108.15 per acre. Charges for property taxes (say $20 per acre) and maintenance (say $3 for fence row weed control) would bring the annual cost of using the land to $131.15 per acre. If the value of the land in beginning-of-year dollars was $2,100, it should be adjusted to the end of the year by the inflation rate. If the inflation rate was 3% per year this would give an endof-year value of $2,163 and an implied opportunity cost of $108.15.

When market sales of land that represent agricultural use value are not available, a more complex procedure is required to obtain reasonable estimates of agricultural use cost. This means that it is necessary to accurately assess land use in the areas by identifying the proportion of total land in farms that is used for production enterprises and identifying the proportion of production acreage devoted to each production enterprise.

A relevant current period must be specified for defining yields and production costs to be used in calculating the annual returns to land for each production enterprise and computing a composite annual rent value. This composite per acre value would be equivalent to a cash rent value and would require no adjustments for taxes or maintenance costs because it represents agricultural returns to ownership. As was the case for alternative, the land use cost is a residual claimant for the owneroperator, even though it is composed of both cash and noncash components.

Mountains and hills cover as much as 93% of Tajik territory. The climate is continental, and the country features a wide range of ecosystems. The four main vegetative zones are desert and steppe in the west, and alpine and meadows in the east. Both the population and agricultural production are concentrated in the valleys in the west of the country. As in all countries dominated by mountains, Tajikistan has a broad range of agricultural regions. Khatlon oblastin the south is characterized by irrigated lowland and has a certain focus on cereal production. Citrus fruits canalso be grown successfully.

The intensive cotton production in the Ferghana valley is an important feature of Sughd oblast. In the regions "under republican subordination," the Hisor valley is a particularly well-developed agricultural region. In this central region access to urban markets is good, land reform is making good progress and the irrigation network is in a comparatively good state. The sparsely populated Gorno-Badakshon Autonomous Oblast in Pamir has very little arable land. Due to the steep slopes, this arable land is frequently terraced. The abandoned highland steppe of Eastern Pamir, where the occasional sheep and yaks graze, is a huge contrast to the intensive, irrigated agriculture in the populated valleys of Sughd and Khatlon oblasts, and the regions "under republican subordination."

Organization of production

The agricultural sector is very important in terms of its contribution to GDP - 24.4% in 2002. The sector employs 65% of the labour force (132% of 1991 levels) and generates 11% of export revenues. Almost three quarters of the population live in the countryside and the importance of agriculture for their survival cannot be over estimated. About 4.1 million ha is agricultural land; the bulk of this land is pasture. The acreage of arable and irrigated land per capita is lower in Tajikistan than in other Central Asian countries. This is a bottleneck in the development of the country's economy, in particular in view of the rapid increase of the population.

Of Tajikistan's total 851,000 ha of arable land, 719,000 ha are irrigated. Irrigated land generates more than 80% of agricultural production, and 85% of water use is for irrigation. Most of the irrigated land, 83%, is found in the Khatlon and Sughd oblasts. There are government plans to increase the irrigated acreage to 1,060,000 ha by 2010 and to 1,200,000 by 2025. However, funding restrictions will make it difficult if not impossible to reach these objectives. The geography and the topography make the development of irrigation more expensive in Tajikistan than in other countries of Central Asia.

For different reasons some of the arable land is not used, including 20-30% of irrigated land. Expensive inputs and deteriorating irrigation systems are important reasons. Agricultural production relies mainly on manual labour, since the material and technical bases for production have deteriorated significantly since the beginning of the 1990s and are not being renewed. Until the dissolution of the Soviet Union in 1991, Tajikistan was an important supplier of products such as vegetables and citrus and other fruits to other Soviet republics. The crops were grown on State or collective farms using industrial farming techniques, but farm workers also grew products for export on private plots. Following independence in 1991, Tajikistan's links with other Soviet republics were abruptly cut. Production of all major products fell drastically, and production is still below 50% of its 1991 level.

Basic food such as wheat was no longer delivered from other parts of the former Soviet Union. These changes affected the economy of the country and the living conditions of its population negatively. With the civil war as an additional negative factor, hunger became a recurring problem. Consumption of meat and vegetables has fallen drastically, while consumption of cheaper food such as potato and grain has increased. Food dominates household expenditure.

A series of droughts in 1999-2001 has contributed to the domestic food shortage, denting a trend of increasing production over the past few years. The country has a 20% food deficit, and even though grain production has increased, the need for imports was above 600,000 tons in 2002. Some of these quantities are supplied in food aid. Chronic malnutrition is still common among children.

The rural population relies on subsistence production on household plots, providing an estimated 50% of total household income in kind and in cash. Nearly all households have access to a small plot (0.1-0.3 ha). In 1996-1998, 75,000 ha were distributed in small plots according to a presidential decree. Total acreage of household plots is 125,000 ha, 60% of which are being irrigated. Most of the households own livestock.

The structure of production changed considerably after independence, as the share of production devoted to basic foodstuffs such as grain and potatoes has increased. However, cotton remains the dominating cash crop. There are still some marginal exports of tobacco, silk, fruits and vegetables to the Russian Federation and other countries. The domestic market is small, and can absorb only restricted amounts, in particular of fruits and vegetables.

The process of handing over to farmers the right to use agricultural land is ongoing, and private farmers now cultivate much of the agricultural land including pasture. In March 2004 there were 20,743 private, dekhkan farms with 923,000 shareholders covering 53% of the land. Some State farms are excluded from privatization, but otherwise the aim is to privatize agricultural production before the end of 2005.

There are three kinds of private farms: farms of individual entrepreneurs, family farms and cooperative ("company") farms, this last category being in some cases only cosmetic transformations of former kolkhozes and sovkhozes. Central and local authorities tend to favour larger farms in an effort to promote cotton production. The "individual farms" and "family farms" are generally very small and may in the longer term be difficult to run efficiently. A typical family farm is 1-4 ha. As the right to land use is being privatized, leasing of land is becoming more common. The leasing of land can be either formal or, as is often the case, on the basis of an informal agreement. There are, however, restrictions on the leasing of land.

Land privatization seems to be having a positive effect on production, but the new farmers are facing a number of difficulties. They do not seem to sufficiently understand their legal rights. Land privatization has not been coordinated with the privatization of livestock, machinery and other means for production.

The new family farms lack capital and machinery. Few can afford to buy a tractor. There is a need to develop cooperation among the new family farms on practical aspects such as the cooperative use of machinery, transport, processing and sale of products, and maintenance of irrigation systems. Some of the previous machine stations are now in private hands and can supply services to the farmers that have the means to pay for help with harvesting and other activities. Water user associations are being established, but this process will take some time.

Markets for agricultural outputs and inputs are liberalized but are developing slowly. With the exception of cotton, the processing of agricultural products is a serious hurdle. The availability of inputs and the further development of markets differ significantly between different parts of the country. Loan agreements often stipulate disadvantageous prices of inputs ahead of the growing season. Restricted access to reasonable credits is also a problem. Cotton farmers can indeed get loans or input deliveries against futures contracts for their crop, but these loans are often expensive. The development of extension services is an obligatory response to the greatly broadened scope of responsibilities of the new private farmer. Some steps have been taken, but the activities do not respond fully to the needs.

Finally, to reach sustainable development, the rural sector requires a road network and communication infrastructure. Transporting products and inputs, even within the country, is today difficult and expensive. Several, mostly small-scale, donor programmes aim at overcoming different obstacles in agricultural production such as supplying micro credit, developing extension services, delivering inputs, establishing associations, rehabilitating the irrigation infrastructure, and seed multiplication.

Production

Cotton

Cotton is at the core of Tajikistan's agricultural production and policy. It brings in much-needed export and tax revenue for the State, and is important as a source of income beyond the farming population. A third of employees in the farming sector depend on cotton. Tajikistan has a tradition and the natural conditions (water, climate) for producing high-quality cotton. Considering the distance from Tajikistan to export markets and the undeveloped transport infrastructure, cotton is among the few possible export products from agriculture in the short to medium term. Cotton is easy to transport and the contacts with buyers on the world market are well developed. It is the second largest export earner for Tajikistan after aluminium, generating about 10% of total earnings. Cotton is cultivated on about 45% of the irrigated land. During the past years annual production has been 350,000-450,000 tons, with a yield generally lower than during Soviet times.

The interest of the State and other stakeholders in cotton is reflected in the continuing top-down, Soviet-style control even if the private farmer is in principle free to decide what to grow. There is still a detailed plan for cotton deliveries and acreages, for the timing of sowing and harvesting, for all suppliers, down to individual farmers. The local administration applies strong pressure to achieve the objectives set higher up. Credits may be conditional on cotton planting. The profitability of cotton cultivation is low. Successful cotton cultivation depends on the application of substantial quantities of inputs, and an expensive credit system has led to a widely distributed rolling-over debt of farmers.

Local monopolies of the privatized cotton processing plants and export monopolies have kept the prices paid to the farmers for raw cotton down. Low world market prices and droughts in recent years have added to these problems. It is difficult to predict the future opportunities and profitability of cotton production in Tajikistan. There are good opportunities for increasing the yield, but world market prices are volatile. An important condition for the sound development of cotton production is that farmers should receive a higher proportion of the world market prices and that the improved profits generated should be used to develop production, and in particular the infrastructure for irrigation.

Monoculture of cotton has negative environmental consequences, and lowers soil fertility. The pesticide and fertilizer use on cotton is high and its cultivation is negative for the humus content of the soil. It is important to improve crop rotation.

Other crops

With a favourable climate and diverse regions, there is a potential for the production of a wide range of crops. Cereals - mainly wheat and barley - are the dominating group of crops besides cotton. The acreage for cereals in 2002 was 364,000 ha. Production increased significantly between 1995 and 1997 but has since stagnated. About half of the wheat acreage is irrigated, and half rain-fed. Crop production on rain-fedarable land has increased considerably during the past years, and the unregulated use of frequently steeply sloping pasture as arable land is increasing erosion. Where the fields are difficult to drain, small-scale rice production is an alternative. There is a great need for animal feed, but the cultivation of forage crops has decreased since Soviet times.

Potato is becoming more important for subsistence and as a cash crop. Production almost trebled from 108,000 tons in 1996 to 318,000 tons in 2001. During Soviet times Tajikistan was renowned for its production of fruits including citrus fruits in the Khatlono blast. There is considerable scope to develop fruit and vegetable production, but it is not possible to do that beyond production for the local market without considerable investments in, among other things, transport and storage systems. Exports to or via neighbouring countries are difficult due to tariffs and other barriers.

Despite the problems, there is a trend towards higher-value products. Onions, potatoes, tomatoes, carrots and strawberries are widely cultivated on household plots, but are mainly sold on local markets. In some regions known for certain products, such as tomatoes and apricots in northern Tajikistan, there is some larger-scale production. A low level of technology, low levels of inputs and low-quality seed material are negative factors for crop production. For example, cereals are often not sown into the sow bed, but instead spread and harrowed into the soil.

Animal production

With the collapse of the previous economic system and the livestock-processing facilities, production and consumption of meat and dairy products has changed significantly. Overall animal production decreased by as much as 85% between 1990 and 1997. During those very difficult times, animal stocks were frequently sold for cash or slaughtered. Cows are particularly important, and if a family does not have a cow it is seen as a sign of poverty. In the highlands of Eastern Pamir, sheep, horse and also yak are the key domestic animals in the difficult struggle to make a living.

After 1997 production recovered somewhat, and the stocks are increasing. Production, however, is still only at 30% compared to the 1980s. Pork production has been almost wiped out in response to national traditions and religious practices today. About 85% of the livestock is now in private hands, and most of there maining State-owned herds will also be privatized.

The large production complexes, or production units such as yak kolkhozes in the Pamir, that were typical in the Soviet period, no longer exist. The herds have in most cases been distributed to individual farmers. Animal production is inefficient. Insufficient quantities of feed, little emphasis on genetic improvement and low quality of veterinary services are some of the reasons. Brucellosis is an example of a widespread livestock disease which can also affect humans. About 3.7 million ha are used as pasture, most of which are summer pasture at high altitude.

Summer pasture is not very intensively used, as the private owners of livestock cannot afford or make the effort to move their small herds. Before the dissolution of the Soviet Union, Tajik farms used summer pastures in Kyrgyzstan. This pasture is not available at present. From the time of the civil war until a couple of years ago, high-altitude summer pasture was not used. As a result much of the winter pasture has been overused with negative consequences for production and land degradation. State-owned livestock is now sometimes moved to higher altitudes in the summer.

Environmental concerns in agriculture

Erosion and desertification

Soil erosion and landslides are significant problems. Erosion is a widespread natural phenomenon due to the relief and climate of the country, but it is accelerated by poor land management practices, such as the cultivation of land on steep slopes; excessive cuttings of forests, shrubs and bushes including wind shelters; over grazing; and improper irrigation. It is estimated that 82.3% of all land and 97.9%

of agricultural land suffers some level of erosion. A new and potentially very difficult erosion problem is caused by the unregulated ploughing of up to 100,000ha of steeply sloping pasture as a way for farmers to increase the restricted acreages of arable land. For example, such fields are common on the hills surrounding the Hisor valley.

In the highlands of Eastern Pamir, one of the few adapted species, the shrub Eurotia ceratoides plays an important role for soil stabilization and also for grazing animals. The plant is used increasingly as firewood, leading to erosion and desertification, but also decreasing feeding opportunities for livestock and wild animals. Erosion affects 60% of the irrigated land. One extreme example is the Yavan valley irrigation scheme. One fifth of this 6,000-ha irrigation scheme that was developed in the 1960s cannot be used because of erosion. Obikiikand Dangara valleys, Urtaboz, Karaul-Tyube and Garauty are other examples of places where irrigated land iseroded.

Increased pressure on the remaining forests by the rural population, compounded by over grazing close to the villages, is a particularly negative factor with regard to erosion. The available winter and summer pasture can feed the livestock, but the farmers cannot transport cattle and sheep to summer pastures. Over grazing is most widespread in Khatlon oblast, but also takes place in Sughd oblast close to villages in the valleys. There are norms for the minimum acreage per grazing animal, but these norms do not seem to be followed. With an increasing animal stock, there is a risk that over grazing will become a more serious problem.

About half the Tajik territory is desert or semi-desert. Deserts are found on the lowland as well as at high altitudes. Erosion and over grazing are two of the factors accelerating desertification. Some others are deforestation, the development of transport and other infrastructure, and mining. It is sometimes claimed that high temperatures and low precipitation during individual years, which contribute to desertification, could be an effect of climate change.

Soil protection activities to decrease erosion and desertification have declined significantly since the 1980s.Similarly, no special measures have been taken in recent years to combat desertification. Planting of forests and other efforts are made mainly at the initiative of the local administration. This is positive, but increased efforts from central authorities would make a difference. Over the past decade only marginal resources have been devoted to monitoring erosion and desertification. The development of monitoring is included in the National Action Programme to Combat Desertification.

Biodiversity of crop plants and domestic animals

Central Asia is the place of origin for many cultivated plants, and Tajikistan is home to a rich spectrum of wild relatives of crop plants. There are also surviving land races of domestic animals. During the Soviet period many of the crop land races were exchanged for a few introduced varieties, but recent expeditions have shown that interesting land races are still available. Much of what has been collected in terms of agro-biodiversity in Tajikistan is now in Russian research institutes. Fruit genotype collections set up during the Soviet period are deteriorating.

During the past few years, expeditions have been organized to collect land races as well as wild close relatives of agricultural crops such as cereals, fruits and vegetables. The objective is to safeguard important genetic material for future generations. It is somewhat surprising that such expeditions have not taken place since the famous Russian geneticist N. Vavilov undertook them before the Russian revolution. A small plant genetic resources unit has been established and a modest seed storage facility is being constructed with donor money. A gene bank for land races of domestic animals has been established with the support of FAO.

Irrigation and drainage

Irrigation and drainage are important for an efficient agriculture, but can also have negative effects on the environment and soil fertility. The link between irrigation and erosion was highlighted in a previous section. During the 1990s the infrastructure for irrigation and drainage deteriorated seriously as the funding for maintenance dwindled from US$ 72 million in 1991 to US$ 6.5 million in 2002. Pumping is an important feature of Tajik irrigation and drainage, and failing pumps and increasing electricity costs are causing problems. As much as a sixth of irrigated land - 130,000 ha - cannot be used. The deterioration of the irrigation infrastructure is slowing down, but significant resources are needed to make real progress. UNDP has calculated that it would cost US$ 106.2 million to rehabilitate the irrigated land that is not included in ongoing World Bank projects. Operating and management costs after rehabilitation are estimated at US$ 21.6 million annually.

Tajikistan has a soil salinization problem that is related to irrigation, even if this problem is less acute than in the downstream countries of Central Asia. More than 15% of irrigated land is salinized and 2,500 ha are lost annually, mostly in Khatlon oblast. Low-quality management, non-existent or non-working drainage systems, and a deteriorating infrastructure are the main causes of salinization as well as water logging. It is claimed that the salt transported by the wind from the dry bottom of the Aral Sea also contributes to salinization. Salinization and water logging are most frequent in Sughd and Khatlon oblasts.

According to estimates by the World Food Programme, almost 40,000 ha of arable land is severely or moderately salinized. More than 40,000 ha are waterlogged. These figures may not give a full picture of the situation, as the resources available for monitoring are restricted, but it is obvious that economic and environmental consequences are significant. The Ministry of Land Reclamation and Water Resources estimates that salinization depresses cotton production by 100,000 tons a year.

Use of fertilizers and pesticides

Pesticide use was high in the 1970s and 1980s, especially for cotton production. In 1985, pesticide deliveries exceeded 18,800 tons. Organo chlorinated compounds could at that time be detected on a considerable proportion of the acreage, and it is claimed that traces of these compounds are still found in the environment. It is unlikely that these traces have significant environmental and health effects, but this cannot be stated with certainty without closer investigation. In the 1990s pesticide use declined dramatically, and during the civil war, restricted access to pesticides led to significant pest and weed problems. The use of pesticides increased after the civil war, but, because farmers are poor, it has stabilized at a low level, less than 2,000 tons per year.

The lack both of training and of information for farmers hampers the good selection and proper use of pesticides. Cheap but inefficient pesticides are often applied. Moreover, pesticides prayers are outdated and, if at all available, do not lend themselves to an even distribution of the active substance. The same applies to fertilizers. The farmers spraying pesticides do not use any protective equipment.

It is noteworthy that, already during the Soviet period, Tajikistan was well advanced in the use of biological and integrated methods to fight crop pests. Air spraying was abandoned early. This helped decrease the use of pesticides in comparison with other parts of the Soviet Union. Today it is the official policy to develop integrated and biological pest management, but the necessary information service and production of biological agents for pest management are not in place. As with pesticides, the application of fertilizers is very low.

Over 600,000 tons were applied annually during the last years of the Soviet period, in comparison with 100,000 tons at present. The use of fertilizers on cereals is estimated to be 10-15% of recommended levels. Most inputs are used on the cotton crop and in the cotton contracts there are provisions for deliveries of pesticides and fertilizers. Fertilizer and pesticide run-off from the fields should be significantly lower, even if increased erosion in irrigated and non-irrigated agriculture is a negative factor. The conditions for the storage, transport and application of the chemicals, however, are inadequate and may cause pollution. Nitrate contamination in the Syr Darya from the cotton-growing districts is still considerable. This problem seems to be less pronounced in the south of the country. For pesticides as well as fertilizers, the lack of monitoring makes it difficult to define the magnitude of the problem.

Pollution from animal production

Most of the large animal production units have been dismantled, leading to a decrease in pollution from manure in some districts. However, manure handling in smaller-scale animal production also pollutes water with nutrients and organic substances, probably with significant negative effects on drinking water. Producers tend to store manure in the villages or in the field for a considerable time before spreading it, which increases the risk of run-off of nutrients and other substances.

Soil contamination and destruction

The state of the waste management facilities causes local soil contamination and problems for agriculture. For example, it is claimed that buffer zones around radioactive waste dumps in Faizabad raion are used by the local population for livestock grazing. Deposition of air pollutants may also cause contamination. The soil is contaminated with fluorine in the vicinity of the aluminium smelter of Tursunzade - on both sides of the border with Uzbekistan - and this is reportedly making livestock ill.

The State Environment Programme for 1998-2008 highlights the importance of land recultivation, anti-erosion activities and improved drainage. Land use is among the priorities of the State Committee for Environmental Protection and Forestry. The central policy document for the agricultural sector, the Mid-term Programme to Overcome the Crisis in the Agricultural Sector and Priority Directions for its Strategic Development until 2005, was adopted in 2000. One of its top priorities is to increase productivity to the levels of the early 1990s, and thus to improve food security for the population. It also aims to improve the quality of Tajikistan's food products and develop markets for them. Privatization of agricultural production before the end of 2005 is a main component of the Programme.

The Programme for the Development of the Cotton Sector for 2002-2005, adopted in 2002, reflects the efforts of the authorities, by using directives and orders in a Soviet tradition, to increase cotton production. This Programme focuses on new technologies, improved and increased deliveries of inputs, the organization of machine stations and new, improved cultivars. From an environmental and land management perspective, it is positive that the Programme includes efforts to improve the irrigation infrastructure, crop rotation, integrated pest control, and control over the import and use of pesticides. Its implementation, however, is weak. Land improvement and land stabilization are important parts of official policy, but there is very little funding available from the national budget. There are no broad programmes for land improvement, even if smaller-scale efforts are made, for example, to plant trees on the raion and local levels. The funding for this comes from different sources, including local budgets.

Tajikistan is a Party to the United Nations Convention to Combat Desertification. The National Action Programme to Combat Desertification was adopted in 2001, but implementation has not yet begun. A GEF project to support its implementation is under preparation. The National Strategy and Action Plan on the Conservation and Sustainable Use of Biodiversity includes a number of actions on plant and animal genetic resources in agriculture, but the lack of funding has been holding back implementation.

The Ministry of Land Reclamation and Water Resources is responsible for managing water resources and the water infrastructure. The Commission for Chemical Safety is responsible for the registration and use of pesticides and fertilizers. Cooperation and coordination among the different authorities is not always smooth. In this context, the cooperation between the Ministry of Agriculture, the State Committee for Land Administration and the State Committee for Environmental Protection and Forestry could be improved. For example, the flow of information among the authorities is limited. There is also considerable scope for improving cooperation and the division of labour between the inspectorates of the two State Committees. Various institutes of the Academy of Sciences and Academy of Agricultural Sciences have a considerable capacity for research. As in other former Soviet republics, the system for research and education is under severe pressure due to a lack of funding. The Soil Science Research Institute and the Institute of Hydraulic Engineering and Land Reclamation, for example, are key research institutes for important issues such as the management of soils and irrigation. The research organizations are in some cases involved in information and training activities for the new private farms.

The Agrarian University in Dushanbe educates students in the whole range of agricultural specialties. It enrols 700 day students each year. A new specialty, agro-ecology, was started this year. There is an obligatory course in environmental protection.

The Agri business Faculty trains future farmers. The two above-mentioned World Bank projects on the privatization and rehabilitation of the water infrastructure for irrigation include a component on extension services. A centre for extension services has been set up. It organizes training courses for farmers, local administrators and specialists. In 2004 the plans are to train 7,000 people. Environmental issues are raised in the Centre's 1- to 2-week courses. The national Association of Farmers was established in 1997 and is developing dynamically. The Association has some 20,000 members using about half of the arable land in 245 local associations. Its main objective is to support the rights of the newly established farmers.

Environmental awareness of farmers and organic farming

The social and economic crisis that ensued after independence and as a result of the civil war can explain why farmers pay little attention to environmental issues, even though, for example, erosion is a direct threat to future production and farmers' incomes. Because of the difficult day-to-day situation, farmers may not even apply their knowledge and experience. If there are no affordable energy sources other than firewood, even forests planted to protect against erosion are likely to be cut down. In addition to the difficult economic conditions, the lack of extension services and of information to farmers seriously hampers the development of agricultural practices that use inputs efficiently and safely, and land and soil in a sustainable way. Few of the new farmers have any agricultural education or access to advisory service seven if the Government and different organizations are making serious efforts. Projects implemented by FAO, other international organizations and bilateral donors also focus on improving the supply of information and training to farmers.

In the present situation, the development of organic agriculture with certification of production cannot be among the first priorities and it is not likely that an internal market for organic products can be developed. However, there is experience in the biological treatment of pests and integrated pest management, and this could be a basis to build on for the longer term. Low labour costs are also a positive factor for minimizing the use of herbicides. Organic cotton production, which is developing in the United States of America, might be an option in the longer term.

Agricultural production in Tajikistan is in a difficult situation, and the sector has to adapt to a dramatically changed structure of production, and with the exception of cotton, new markets. At the same time there are only marginal resources available to adapt to the new situation. It is understandable why environmental issues have not been a primary focus of this battered sector. In the current situation, it is very difficult to introduce elaborate environmental protection schemes in the agricultural sector. New or changed practices are likely to be introduced successfully only if they also improve production and living standards in the countryside. It is important to support the general development of the agricultural sector.

The skills and financial situation of the new farmers and farm managers need to be improved, and markets for inputs and produce developed. Significant efforts are being made, with the loan-financed projects to rehabilitate irrigation infrastructure probably being the most important. A more efficient agriculture with an improved economy is likely to have a positive impact on soil and land management, on the efficiency of the use of inputs, and on the status of the water infrastructure.

Extension services and training are key instruments in the development of private agriculture. An important task for the extension services is to promote agricultural practices that decrease erosion, and ensure the safe and efficient use of pesticides and fertilizers. The attempts to develop extension services are commendable, but the lack of funding and a general mistrust towards the authorities are drawbacks. The scientific basis for extension services is important. Agricultural science is reasonably well developed, but its funding is very limited.

3

Agricultural Geography: An Overview

Around ten to twelve thousand years ago, human began to domesticate plants and animals for food. Before this first agricultural revolution, people relied on hunting and gathering to obtain food supplies. While there are still groups of hunters and gatherers in the world, most societies have switched to agriculture.

The beginnings of agriculture did not just occur in one place but appeared almost simultaneously around the world, possibly through trail and error with different plants and animals or by long term experimentation. Between the first agricultural revolution thousands of years ago and the 17th century, agriculture remained pretty much the same.

In the seventeenth century, a second agricultural revolution took place which increased efficiency of production as well as distribution which allowed more people to move to the cities as the industrial revolution got under way. The 18th century's Europe became sources of raw agricultural and mineral products for the industrializing nations.

Now, many of the countries which were once colonies of Europe, especially those in Central America, are still heavily involved in the same types of agricultural production as they were hundreds of years ago. Farming in the twentieth century has become highly technological in more developed nations with geographical technologies like GIS, GPS, and remote sensing while less developed nations continue with practices which are similar to those developed after the first agricultural revolution, thousands of years ago.

About 45% of the world's population makes their living through agriculture. The proportion of the population involved in agriculture ranges from about 2% in the United States to about 80% in some parts of Asia and Africa. There are two types of agriculture, subsistence and commercial.

There are millions of subsistence farmers in the world, those who produce only enough crops to feed their families.

Many subsistence farmers use the slash and burn or swidden agricultural method. It's a technique used by about 150 to 200 million people, and is especially prevalent in Africa, Latin America, and Southeast Asia. A portion of land is cleared and burned to provide at least one and up to three years of good crops for that portion of land. Once the land can no longer be utilized, a new patch of ground is slashed and burnt for another round of crops. This is not a neat or well-organized method of agricultural production by it is effective for farmers who don't know much about irrigation, soil, and fertilization.

The second type of agriculture is commercial agriculture, where the primary purpose is to sell one's product at market. This takes place throughout the world and includes major fruit plantations in Central America as well as huge agribusiness wheat farms in the Midwestern United States.

Geographers commonly identify two major "belts" of crops in the U.S. The wheat belt is identified as crossing the Dakotas, Nebraska, Kansas, and Oklahoma. Corn, which is primarily grown to feed livestock, reaches from southern Minnesota, across Iowa, Illinois, Indiana, and Ohio.

J.H. Von Thünen developed a model in 1826 for the agricultural use of land. It has been utilized by geographers since that time. His theory stated that the more perishable and heavier products would be grown closer to urban areas. By looking at the crops grown within metropolitan areas in the U.S., we can see that his theory still holds true. It is very common for perishable vegetables and fruits to be grown within metropolitan areas while less-perishable grain is predominantly produced in non-metropolitan counties.

Agriculture has always been a critical determinant of human geography and community development. Population concentrations historically located in either the most inaccessible or accessible places. Inaccessible sites were chosen for the security they provided. Accessible sites were chosen for proximity to food- where the land, water, and climate most favored farming. Mumford says that the "first germ" of a city is in the ceremonial meeting place. Fixed investments were also made in places that served as trading points or storage locations. Permanent settlements associated with agriculture appeared in the Mesolithic period, about fifteen thousand years ago.

Agricultural Geography and Rural Development

In modern times, sites central to agricultural regions still attract concentrations of population, as transshipment, storage, and food and feed processing, and marketing activities optimally locate in the middle of the farm supply region. In modern as in Neolithic times, the establishment of an agro-industry in the most accessible center of the growing region gives rise to a point mass of population there. The center's optimality is reinforced if it also becomes a transshipment point. And, the center remains an optimal site for agroindustry regardless of who is responsible for farm product transport; relative transport cost rates, or industrial input:output rates. The relatively dense place where offfarm activity locates becomes an urban concentration. Thus, because it locates to minimize the cost of obtaining inputs from a supply area, most agroindustry is urban even though it is materialsoriented. In sum, agriculture historically determined the locations of cities. Agriculture also used to determine the sizes of cities. Hunting and gathering sustains less than ten people per square mile.

Aristotle thought that the ideal city should contain not more than 5,000 citizens. Until the twentieth century, the extent of urbanization was limited by the food surplus available to the people who lived in cities. Thus, until very recently, almost everyone lived in what are now called "rural communities." Before the industrial revolution only about ten percent of a region's population could be sustained in cities, and even the largest cities were generally concentrations of fewer than 150,000 people. Rome reached a peak population of about one million inhabitants, but shrunk to 100,000 during the Middle Ages.

As summarized in Table 1, Asia was more urbanized than all other regions. That was due to Asia's agriculture. Rice can be harvested up to three times per year and provides more calories per hectare than wheat. Traditional Asian rice farm lands could support five times as many city people per farmed area as European or African wheat farms.

So Asian cities could be much larger. Also until the industrial revolution, farms followed people. In 1824 Von Thunen wrote that when land is allocated to its most profitable uses, the activities that generate the highest revenues over production and transport costs (bid rent) will be found at each location.

Table 1. Percentage of Population Living in Urban Areas

	1300	*1500*	*1700*	*1800*	*1900*	*1950*	*1980*	*2000*
World	9	9	10	9	16	26	37	47
Africa	4	5	4	4	6	12	25	37
America	8	8	11	12	29	48	64	76
Asia	11	12	11	9	9	15	26	37
Europe	10	10	11	10	30	43	64	74

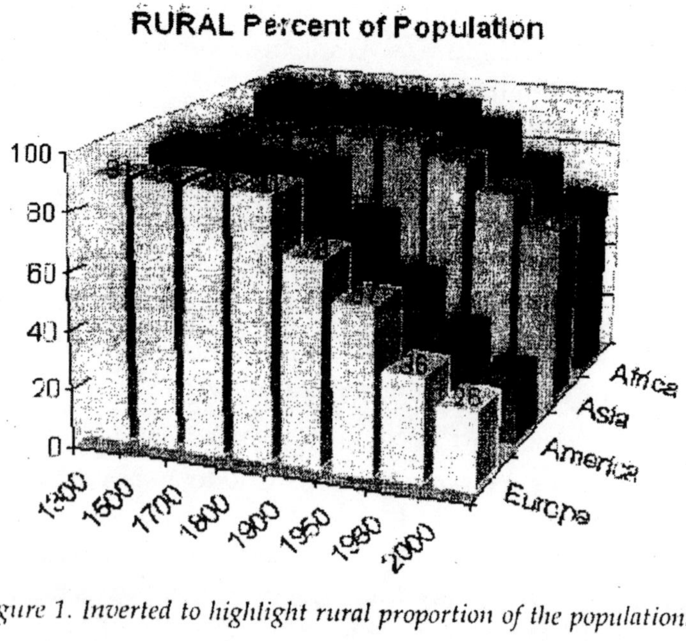

Figure 1. Inverted to highlight rural proportion of the populations

Von Thunen rationalized the observed geography of farm land use around towns: close-in intensive vegetable cropping and milk production, then extensive field crops and grains, and range lands farthest out. His insights explain urban as well as rural land use patterns today. In this way, cities have also determined the location of agriculture. The extent of farming around a market (or city) is limited by the cost of transporting the land's produce.

Let π denote the profits of an establishment in a market-oriented industry at distances (m) miles from the market:

$$\pi = P \cdot Q - c \cdot Q - t \cdot m \cdot Q - R(m) \qquad (1)$$

where P and Q denote output price and output quantity; c is production cost per unit (exclusive of the fixed cost of land rent); t is the output transport cost rate per unit per mile, and R(m) denotes land rent at location m. The extensive margin (m) is the distance at which bid rents are exceeded by opportunity costs. If there are no alternative land uses, this is where rents are zero: $m = (P\text{-}C)/t$. Using the same basic analytical elements, we can easily show that where transport costs to markets are high, the payoff to production for market is low, so farm productivity remains low. Assume, for example, a "textbook" case where total costs (per farmer or per acre) include fixed costs and a cubic component (implying quadratic (U-shaped) marginal cost curves): $C = f + Q^3$. Then, profit = $(P - t \cdot d)Q - f - Q^3$, and the optimal output (per acre or farmer) is found from the first-order condition: $P - t \cdot d - 3Q^2 = 0$, to be $Q^* = [(P - tod)/3]^{1/2}$.

Taking the derivative with respect to transport costs or distance (applying the chain rule) shows that optimal output per farmer or per acre is decreasing and strictly concave in t and d. In other words, farming for market is not rational too far from the city, where "too far" is determined by transport costs as well as distance. This constraint continues to bind in developing countries today. For example, Omamo applied an agricultural household model with explicit transport costs to investigate the rationality of seemingly inefficient cropping choices. Households in a remote district of Kenya regularly devote larger shares of resources to low-yielding food crops than they do to cash crops that have higher market returns. He concluded that "improved rural road networks that reduce [transport] costs could abate motives to meet food needs through domestic production and promote specialization that raises farm incomes."

The implication of the transport cost constraint on the profitability of farming in developed countries is different. Where credit markets function properly, the transport cost rate (t) is clearly endogenous, through the shipper's choice of transport mode and scale. For example, in Iowa farmers own semis so they can ship large quantities of product across states. McCann shows that under very general conditions the optimum size of the vehicle increases with the haulage distance and haulage weight. This means that transport cost rates decline with distance, and that profits are more likely to be strictly concave in distance between input and market locations, reinforcing and the endpoint optimality condition that rationalizes establishment clustering at input or market locations- never in-between.

Both farming and cities were transformed during the eighteenth and nineteenth centuries. A revolution in agricultural production preceded the industrial revolution. Farmers in Holland and Belgium (and England) developed intensive methods that dramatically increased wheat yields per hectare. This allowed for the release of labor from farming, into other occupations, and supported the expansion of urbanization and industrialization. Among the new machines were new modes of transportation.

The shift from animal power and transport released acreage from the production of forage and feed for draft animals, so that all crop land could be devoted to food production. Dramatically increased food production per acre and dramatically reduced transport costs meant that much larger concentrations of people could be supported in cities very far away from farms. By 1850, London was the largest city in the world, until 1950, when it was outranked by New York, which was outranked in 1980 by Tokyo. As made clear in equation (1), every innovation and invention that increased productivity (reduced the costs of production) or reduced transport cost rates, allowed the farming activities to be profitable farther and farther away from the city. Modern people no longer need to live near farms. And farms no longer have to be near people.

Electrification and the development of telecommunications and passenger elevators supported an exponential rate of urban concentration. "Before the advent of the elevator, the height of buildings had been limited by human endurance in stair climbing. Urban buildings were four to six stories...but with the introduction of elevators, buildings could be made as tall as construction techniques and engineering would allow." Through rapid, vertical construction, the population of Chicago grew 75 times in forty years from 20 thousand in 1850 to 1.5 million by 1890. (=11%/yr) The increases in market size made possible by transport cost reductions also support 'Adam Smith' localization economies of scale: the division of labor is limited by the extent of the market.

As the market expands, a preexisting plant can enjoy profits due to internal scale economies. That profitability entices new establishments in the same industry to open in the same location. As the number of plants rises, the local demand for inputs into the industry rises. At some point, the market for inputs becomes large enough to support plants that specialize in the inputs. This outsourcing allows both the upstream input supplying plants and downstream industry plants to profit from both the economies of scale and the productivity gains due to specialization. Thus, in larger markets, both costs and prices are lower. See Duranton (1998) for an excellent abstract model of this process of positive feedback and urbanization.

Indeed, measured labor productivity is higher in markets large enough to support multiple establishments in the same industry. The preoccupation in the current literature is to attribute this to the claim that urban workers are more productive than rural workers because of agglomeration economies and spillovers. A much more compelling argument is that while all workers and firms may be drawn from the same distribution of productivity, the competition in larger markets weeds out the higher-cost/lower productivity establishments. Thus, the observed distribution in cities is truncated from below, leading to higher average measured productivity in larger markets that can support many firms, than in small towns where there is only one. Are bad roads good for rural development?

Since the 1890s, overland transport costs have fallen 90%. And just since the 1950s, water-borne transport rates for bulk products have decreased up to 70% due to improved maritime technology. Some argues that transportation remains a major determinant of location because relative spending on transport (6-8% of GDP across

OECD countries) has been relatively constant during the industrialisation and modernisation process. In contrast, Some argue that transport costs for goods have fallen enough to be negligible. Their two positions are reconciled by noting that while transport cost rates have fallen, our use of transport cost services has increased, and the mix of modes of transport has changed substantially. In particular, in the USA we buy more things that require high cost truck transport (about 25¢ per ton-mile) and much less using cheap rail transport (3¢).

Water-borne transport has always been a hundred times cheaper than overland transport, and rail transport about ten times cheaper than truck. The improvements in transport technology and infrastructure over the past century have meant that the relative costs of moving things within countries has fallen more than the cost of moving things between countries. What this has meant for regional and rural development depends in part on how it has affected the size of markets and thus the scale of enterprise.

First, consider the effects of the technological changes in transport on world trade. Transport cost reductions do not seem to have been very important in accounting for growth in developed country trade. Baier and Bergstrand's empirical investigation of world trade among several OECD countries concludes that transport-cost declines explain only about 8% of the growth since World War II. Most of the growth in trade (67%) is explained by income growth, and the rest (25%) by tariff-rate reductions. In developing countries, however, overland transport constraints still bind significantly. For example, conclude that the relatively low level of African trade flows is largely due to poor road infrastructure, and confirm that an improvement in infrastructure from the 75th percentile to the median would lower transport costs by 12 percentage points and increase African trade volumes by 28 percent.

Second, consider intraregional trade. A region need not export to enjoy the benefits of returns to scale, specialization and trade if its domestic market is large enough. Frankel, Stein and Wei hypothesized that the increase in welfare stemming from intra-continental trade is larger the higher are the costs of international trade. World Bank economist Azita Amjadi with Alan Winters show empirically that intra-bloc transport costs impede the economic integration of the Americas, and that "reducing such costs could be an important complement to integration."

The point is that reductions in overland transport costs make continental markets more accessible, so that the benefits of trade can be captured just as well through intraregional specialization and intraregional trade. The economic history of the USA is one example of the efficacy of domestic market led growth. China offers another example. Shiue's study of eighteenth century China also shows that domestic market integration there was a significant source of economic growth. And we have already discussed the impact of transport cost reductions on the optimal intensity or productivity of farming.

In developing economies, overland transport cost reductions can help expand the market size for domestic producers. Alternatively, in developed economies where market areas are already allinclusive, reductions in transport costs cannot make all market areas larger. As all transport cost rates fall, larger producers or firms expand their market share at the expense of smaller ones. This is one of the many arguments that lead analysts to conclude that intranational transport cost reductions could actually work against rural development.

To demonstrate, consider two competing suppliers A and B. Assume that A is the lower cost/larger supplier. Formally: $c_B = (1+\beta)c_A$, $\beta>0$. Note that average cost (c) is endogenous; A may operate at lower cost simply because it enjoys internal economies of scale, but we will ignore that here to close the model.

Let the delivered price charged by A just cover unit costs plus transport: $P_A = c_A + t \cdot m_A$, where mA is the distance to a customer from A; and the delivered price of B be $P_B = c_B + t \cdot m_B$. Furthermore, normalizing the distance between A & B = 1; $m_B = (1-m_A)$ so that each firm's market radii are the market shares on the interval AB (assuming customer density is constant for simplicity). Under these abstractions, $P_B = (1+\beta)c_A + t(1-m_A)$.

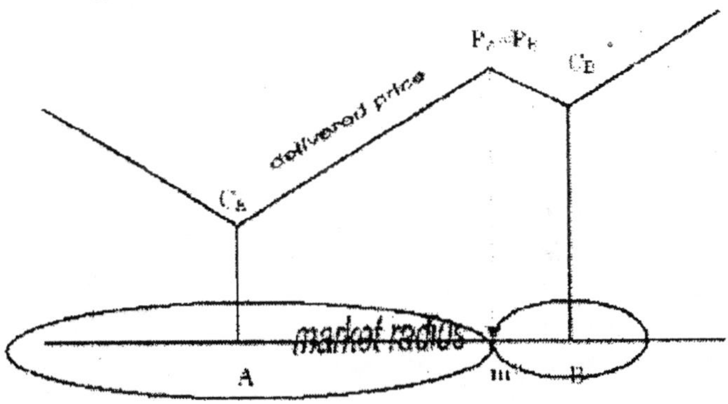

Supplier A's market radius (or market share) is the m* at which A's delivered price equates with B's: $m_A^* = [\beta \cdot c_A + t]/2t$. By the quotient rule,

$$\frac{dm_A^*}{dt} < 0.$$

As transport cost rates rise, A's market share declines. Conversely, as transport cost rates decline for all suppliers, the supplier with the larger initial market area will gain market area (market share) at the expense of the smaller supplier.

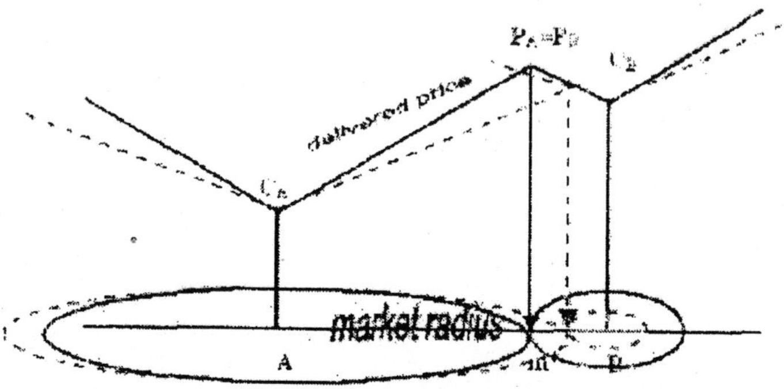

Another argument against transport cost reductions as a rural development strategy arises when some transport cost rates fall relative to others. In particular, using a general equilibrium model in the von Thünen tradition, Nerlove and Sadka show that as the cost of transporting agricultural goods falls relative to the cost of manufactures transport, it becomes economically feasible to cultivate land farther and farther from the city. The relative reduction in agricultural transport cost rates leads to (1) less farm labor per acre, (2) lower agricultural terms of trade, (3) a lower and more dispersed rural population, and (4) an increased proportion of the labor force that works in cities.

In other words, according to their hypotheses, good roads are bad for rural development. Nerlove and Sadka's model, however, does not allow for the endogenous location of non-farm activity. Thus, it cannot be used to explain why there are any non-farm households outside the city, and it is not a model that should be used to analyze rural development in general. By the same token, a spatial agglomeration model designed to explain the increasing regional disparity within China. Given the coast's geographic advantage in international trade, it is the initial location for industrial agglomeration. The predominance coastal cities is reinforced by the positive feedback of external increasing returns to scale. The location disadvantage of the interior is due to their higher transportation cost on international trade. Hu writes, "the model suggests that increasing domestic accessibility can actually make the interior worse off." This pessimistic conclusion may be the consequence of the (dubiously appropriate) assumption that all of China is already within some firm's market area, in which case transport improvements help the big get bigger. But if in fact there are inaccessible places, improving their accessibility will be beneficial.

The most famous demonstrations of transport cost reductions leading to rural depopulation are arguably contained in the recent new economic geography. Drawing on the journal articles they published during the 90s, Krugman, et al. model urbanization relative to 'agriculture' by formalizing the trade-off between increasing returns and the costs of serving a dispersed market of 'farmers' tied to their land. They pose different models by choosing among four sets of assumptions (i) space is discrete (two regions), continuous and bounded (racetrack), or continuous and unbounded (von Thunen-style); (ii) labor is sectorally and/or regionally mobile and/or immobile; (iii) land and rent are ignored or explicit, and (iv) goods and industries are homogeneous or differentiated, final or intermediate, and free or costly to transport.

All their models are closed using a Dixit-Stiglitz formalization of inframarginal monopolistic competition and 'ad hoc dynamics.' Regardless of which set of assumptions they use, all their two-region models with mobile factors generate just two types of stable spatial equilibria. The (non-farm) economic activity subject to increasing returns is either symmetrically dispersed or fully concentrated in one of the regions. As the authors say, they want to explain agglomeration. They are careful to never say anything about rural development or policy. And that is good, because while the empirical evidence is consistent with the implications of the von Thunen model; it is not consistent with the implications of most new economic geography models.

Over the past two hundred years, while the share of the U.S. population that are farmers has fallen from 70% to 2%, and the share that is rural has fallen from 90% to 24%, the share of the population that is rural, but not farmers, has stayed remarkably stable at about 20%. Similarly, in France, although the number of farmers has fallen dramatically, France's non-farm rural population has remained around eight million since 1850. This stylized fact poses quite a puzzle. It also has very interesting implications. It contradicts Krugman-style new economic geography models which find no economic rationale for the existence of any non-farm, remote, low density populations. And it contradicts the claim that agriculture is the long-run economic base of rural communities.

The fact the shares or levels of non-farm rural populations have not fallen with the farm population indicates that the rural nonfarm population of a region does not appear to depend, in the long run, on the number of farmers. Contrary to the assumptions underlying most new economic geography models, an "immobile farm population" is not the main dispersive force in developed economies. A few new economic geographers, however, have succeeded in rationalizing the stability of coexisting small and large concentrations of population and diverse economic activity in one competitive economy. To do so they formalized other dispersive forces. Kilkenny for example, assumes that competition for workers and customers is localized rather than inframarginality (that local wages and market prices are unaffected by the entrance of a new business). Second, she formalizes the decision to invest in a locality according to latent or potential demand for local varieties rather than current expenditures, which are zero on things not yet produced there. Third, people choose locations to maximize real rather than nominal income; where the real income of a rural resident is higher than that of a city resident who earns the same nominal wage, if rural rents are lower.

Fourth, where agglomeration economies lower costs, rents are higher. This is also obvious from Equation (1). A reduction in average production costs, due to rising agglomeration economies of scale, for example, would raise the rents that could be extracted in the city, as well as extend the margin of urban land use. Thus, cost-savings due to agglomeration economies in cities are offset by higher urban land rents. Kilkenny's models show that although transport cost reductions initially support concentration, they ultimately support the expansion of the non-farm rural population. The last point is that if overland transport is affordable and local credit markets function properly, rural communities do not need to depend on export agriculture. They can develop by providing and trading goods and services among themselves.

4

Origin and Spread of Agriculture

Importance of the adoption of agriculture and shift from food collecting to food production is doubtless one of the most crucial events in the Global History. Therefore, from relatively early on questions about origins of such crucial moments are sought. It became one of the great questions of the Prehistory and there is no theoretical framework or school or thought which have not tried to apply their ideas and methods on this very important question. None of the greatest archaeologists of respective times could let the debate over such important question to pass without their views and models, or at least their criticism about the proposed models.

In traditional paradigm, agriculture was often seen as a dominant and evolutionary higher mode of the production, while Hunter-Gatherers were observed as a living past. However, closer investigation of the contemporary Hunter-Gatherers demonstrated highly successful adaptation of such systems as well as traits of complexity previously reserved for the 'evolutionary higher' agriculture. Therefore, these two systems of subsistence must be observed as two frameworks posing constrains and possibilities to a group of humans, not as steps in technological progres and social evolution. Adoption of this notion had profound effect on the more recent theories of agriculture imposing explanations not just from agricultural perspective, but primarily looking from the position Hunter-Gatherer society might have in the transition. It also puts more emphasis on the social organisation and structure in transition to farming.

Almost all explanations of the origins of agriculture assume that, providing familiarity and knowledge of life cycle of potential domesticate or cultigen, and providing the need and appropriate mechanism to do so, forager will make use of domesticates, and eventually start farming. Therefore main objective is to identify those elements and explain them.

Here, we will try to present one of the most usually used ones, and try to point out the main weaknesses as well as strenghts of each model .

Models about Origins of the Agriculture

'Oasis Theory'

One of first sound archaeological explanations (albeit ones that extraterrestials brought seeds of domestic plants...)was put forward by the greatest archaeologist of first half of the 20th century Gordon Childe. In his book published 1928 called 'The most Ancient East' he put forward "Oasis" theory suggesting that Post Pleistocene period was major period of desiccation, and that on the Near East humans and animals seek refuge in fertile circumscribed oases, like the Mesopotamia. This resulted in closer contact between humans and animals. Such availability of potential domesticates resulted in establishment of the agriculture i.e. in the human control over the domesticates. This theory was in 1960 modified by Braidwood who narrowed scope of such oases on the areas of Taurus and Zagros mountains, being first archaeologist to introduce notion of the Fertile Crescent into quest for origins of Agriculture.

"Oasis" hypothesis have several major flaws. First it is not much of the explanation as it does not explain nature of relationship between would be domesticates and humans, if such relationship existed at all. Also, in its basis it is not just descriptive, but it is deterministic as well. Even worse, this environmental determinism is built upon presumption which do not fit into climatic indicators pointing out towards a stabile climate during the period in question. This model obviously grew up out of traditionalistic and diffusionistic view of agriculture as a evolutionary higher step, but on the end not just not being valid, but not being explanation at all.

Ambitious new processualists like Lewis Binford himself in 1968 paper Post Pleistocene Adaptations, slightly modified by Flannery in 1969 , but mainly staying on party line, and demographic works of Cohen and Hassan in 1970s. Realising that Hunter-Gatherers are usually well adapted and therefore have no need to turn to more time consuming agriculture, unless they are forced to, pointed research towards identification of potential incentives and motors of change. Such motor of change was seen in the first place in the population growth and/or environmental change which would trigger the inequilibrium within culture as a system and therefore initiate respond mechanism as adoption of agriculture as a mechanism of restoring equilibrium. Classical example represents Binford's hypothesis of origins of agriculture in the Levant, where after successful marine resources based strategy population grew and overspillt into hinterland, only way to survive was the exploitation of wild grasses and other domesticates and cultigens available locally.

Such hypotheses are clearly product of Processualists close links with system theory, and its view of the society as a system and its perpetual restoration of equilibrium through feedback mechanisms. Therefore, population growth over the capacity of environment will result in lack of food and put system in inequilibrium, which could only be solved with new food production strategies.

Apart from not fitting the actual evidence (Mesolithic in the Levantine was mainly land orientated and the centres of the domestication are in primary, "optimum" areas), such processualist models are being heavily criticised for being monocausal, and giving too much space to deterministic prepositions. Nevertheless it remained quite popular primarily because it deals with elements relatively easy to identify in the archaeological record such are population growth or environmental change.

Therefore, it is not rare encountered in the 1980s, typical one being Zvelebil and Rowley-Conwy explanation of sustained adoption of the agriculture in complex Ertebolle society of Scandinavian Mesolithic. Similar models for the Near East are proposed by McCorriston and Hole and Moore.

Functionalists 'Pull' Model

Essentially similar to this models are next group of models being dubbed "Pull" models. They are not satisfied with one factor compelling foragers to use domestic species, but rather see certain characteristics of potential domesticates making them attractive for human exploitation. This leads to the intensification of their use by the foragers, resulting in greater reliance on such species. That will again result in close relationship and control over life cycle and therefore in gradual genetic change on the side of domesticates and on the other hand it would initiate population growth. However, once population have passed threshold of foraging capacity of the environment, there was no turning back, and farming was started to be practiced, because of its higher yield per square unit.

Such model are devised by Flannery for rise of agriculture in Mesoamerica, where such relationship is seen between human groups and maize and beans. Again, this hypothesis can be linked with strong processual influences, as putting forward shifts in trajectory through positive feedback, however, with less severe monocausality or perils of determinism behind it.

'Social' Model

With increasing interest of Post-Processualism in the context , in the specific, and rather seeing society as a web of relationships not as a machine or a system, as well as in relations of the power within society, new models has appeared. Such models, are often known as 'Social' models or 'competitive feasting' models, explaining change in the society primarily through interactions within society.

The best known models are ones proposed by Bender and Hayden. Both of them seek changes within the Mesolithic societies themselves. In high food stress circumstances foragers will make alliances based on food sharing and in order to prevent depletion of the resources any form of the competition through gifts of food is excluded. However, in areas relatively rich and with numerous r-selected species which are virtually indepletable, sedentary and affluent forager societies are able to form, and food depletion stresses disappear opening opportunities for competition of ambitious individuals for power through gifts of food, resulting in web of loyalty and labour dues. Again this competition will demand more and more resources and in order sustain pace of competition, cultivation of the domesticates started as more efficient way of exploiting the environment, although already known but without real need to start practising it.

Evidence for this model is found in various complex forager societies (from Natufian to Archaic Indians, or Oaxaca culture in Mesoamerica), and reinforced by the facts that centres of domestication were usually in optimal, rich zones or by possibility that primary use of barley and wheat was for brewing. Therefore, we should look for such complex foraging societies as potential 'customers' (Ertebolle, Iron Gorge?), and 'inventors' (Natufian, Oaxaca) for agricultural package .

This model, may have substantial flaws, nevertheless being one of the freshest approaches, is adopted by numerous Post- Processualists, like Hodder himself. The main reason for popularity of such model is in overcoming weaknesses of the most of functionalist's system, main ones being determinism and view of change always coming from the external factors of the society and also enabling for more symbolic processes to came to fore as in Hodder's model of mechanism of domestication through dialectics of notions of the Domus and Agrios.

For the first time, change has been explained as coming from within the society itself, and internal relationships within the society. On the other hand elements of this model of change are far more harder to see in archaeological record, and this fact often detract archaeologists from the model. Nevertheless, further work in the field and developments of theory and practice will certainly modify or reject such view.

Wave of Advance Theory

One of first model was rather simple diffusion assuming absolute dominance and attraction of agriculture. Such model is one proposed by Ammerman and Cavalli-Sforza, based upon notion that agriculture will soon produce population increase substantial enough to fuel diffusion spurred by need for a new land. This produced uniform "Wave of Advance" of farming techniques inevitably sweeping everything in front of him, coming in constant speed of 1km a year. Such pattern was primarily devised through analyisis of C-14 dates for the earliesty farming communities which has obviously fall off pattern going from the Southeast to the Northwest.

Original map of the Ammerman and Cavalli-Sforza showing distribution of C-14 dates across the Europe (years are B.P, and all dates are from excavations before 1971). Arcs indicates predicted progress of the Farming. Dotted lines are only allowance for delaying processes predicated by the Theory. This model is today refused as oversimplification, as looking on the problem strictly from the agriculturalist's perspective, it rules out decision and agency of indigenous foragers and does not actually fit into data, in the first in senese that settlement data from the Near East which does not suggest instant increase of population substantial enough to support

Acculturation and Frontier Theory

This model basically represents modification of the 'Wave of Advance' hypothesis was modified, by putting focus on the circumstances in which foraging group adopted agriculture. Also this view realises adoption of agriculture not as necessity but as choice and that actual process of adoption is not uniform but consisting of three successive phases (availability phase, substitution phase and consolidation phase). This model is chiefly advocated in work of Zvelebil and Rowley Conwy, Alexander and others.

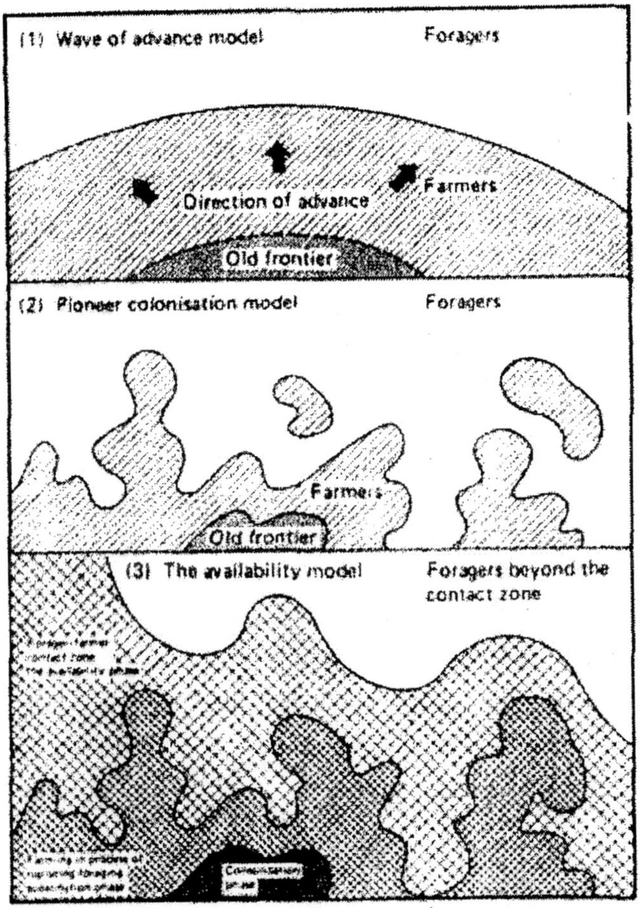

Figure 1. Acculturation and Frontier Model

MODELS ABOUT SPREAD OF THE AGRICULTURE

Spatial models

1.) Wave of Advance (note uniformity of line)

2.) Pioneer colonisation model, not much unlike one on the West of North America in the 19th C.

3) Availability model, note irregularity of frontier (as it depends on the choice of foragers to enter the substitution phase from the availability phase)

However chief weakness remains the explanation of incentives of the change mainly in the 'outside', environmental and population pressures. Nevertheless it is significant as emphasises acculturation and exchange and spread of the ideas as main dispersion mechanism, rather than population displacement, which is actually rarely attested in the archaeological record. Also it has transcended limitations of the 'agricentric' view of the transition putting accent on choices and strategies of the Hunter-Gatherers.

Question of the dispersal mechanisms remains rather open, and new data from the excavations and especially DNA extraction techniques have great potential for resolving the question.

Indigenous Development Therory

Most of the other models presume invention and dispersion from one focus, one region. However school of thought based around Higgs and 'Palaeoeconomy' (including Jarman, Dennell, Barker trend denied dispersal from one area arguing for several parallel focuses of domestication. It put forward study of the actual processes and mechanisms of the domestication, and showe that manipulation of the animals and the plants and interference with the environment on the larger scale has its roots in the pre-Neolithic times not just in the Near East but in the rest of the Europe as well. Therefore evidence from Balkans is often interpreted in terms of indigenous domestication processes (barley, oat, sheep, cattle and pig which all arguably were endemic in the Area), however allowing for processes of dispersal and acculturation, insofar that the key to well organised farming was through adoptions of wheat from the Near East.

Locations of the Transitions to Agriculture

By the Boreal period environment of the Region recovered from the glaciation, becoming not too dissimilar from the one today. Such environment was populated with Forager groups, drawing heavily on the Upper Palaeolithic tool kits. This Epipalaeolithic occupation is evident on sites such are Peres and Backa Palanka in Pannonian plain or Pobiti Kamani in Bulgaria.

The Upland mountain ranges were inhabited as well with local Hunter-Gatherer populations, such as at the sites of Crvena Stijena or Odmut in the Dinarid mountains. Both lowland and upland foragers relied heavily on the hunt of local animal species as well as on gathering of nuts and fruits. Also, the possibility of lowland-upland transhumances in the wake of the deer in the Boreal is sometimes proposed.

Opposed to this Epipalaeolithic Hunter-Gatherer groups were two "Real Mesolithic" groups. One of them is represented in the Franchthi cave, in Greece now on the coast of Sea and the other region is the Danube Gorge, which after intensive survey revealed 11 Mesolithic sedentary sites, the most important being Icoana, Ostruvul Corbului, Ostruvl Banatului, Padina, Vlasac and Lepenski Vir which is the most fascinating.

Regardless of mechanism of development of the farming, the first farming communities became to be visible on the beginning of the Atlantic period. The first farming communities are ascribed to the tell type sites in the

Southern Balkans. The first occupation of the tells in Macedonia, Thessaly and Knossos (for which aceramic phase is proposed as well) starts around 6000 BC, but some prefers lower dates of c.7000B.C. for Crete and Macedonia, 6300 BC for Thessaly and c.6000 for southern Bulgaria, while Central and Northern Balkan early farming communities (Starcevo,Koros and Cris cultures) are usually defined in terms of 6000-5500 BC. On the basis of the pottery styles six cultures were defined:

— Early Neolithic in Greece

— Karanovo I in southern Bulgaria

— Kremikovci overlapping between western Bulgaria and Macedonia

— Starcevo in central Balkans.

— Koros in river network of Danube, Tisza and Koros

— Cris in Romanian part of Koros-Danube network

On the basis of cronological and differences in other areas of inquiries two cultural complexes could be devised. Therefore, first complex would comprise Karanovo-Early Neolithic of Greece-Kremikovici cultures, and could be dubbed

Anatolian-Balkan complex due to broad parallels and similar features to the Anatolian Neolithic. This complex is giving impression of consolidated, fully formed Neolithic, well organised farming economy based upon cultivation of pulses, grains for human diet and other cereal for fodder on well drained, easily worked sandy soils on the lowland-upland contact zone. Some evidence even indicate crop rotation system. Second mainstay of farming were primarily ovicaprines (not rarely making up to 80% of bone assamblage) which probably involved seasonal movement between the summer upland grazing and lowland winter pasture. Percentage of hunting game was negligible. Such farming system had focus on the tell type sites (Yet another parallel with the Near East), representing larger (up to 4.75ha, but mainly around 1-1,5 ha) long term focuses of occupation (tell at Karanovo has accomulated 12 meters of cultural deposits just from the Neolithic to the Bronze age!) and links with the place.

Reconstruction of the Sesklo tell in the Middle Neolithic (5000-4000 BC). By the end of Early Neolithic, tells of the Anatolian-Balkan complex probably looked very much like this one, only probably without the stone circuit around the tell.

This solid farming economy probably formed stabile and successful farming communities with rectangular, often painted houses (again, parallel with Anatolia together with consolidated farming), and long distance trading links and growing complexity. However, now it is clear that this was not implanted from the very beginning of the Neolithic occupation as the earliest levels of the tells shows much more affinity with lifestyle of the first farmers in the Central and Northern Balkans and SE Europe.

Going into the Central Balkans and North of SE Europe, one naturally is moving from dominantly Mediterranean and Sub-Mediterranean Environment into increasingly Continental one. This certainly modified first farming, which have to underwent "Balkanisation".

Whether result of indigenous domestication processes or acculturation, the final product differed from the first communities on the South, and it is almost always seen as the adaptation. Starcevo-Koros-Cris complex gives impression of still mainly mobile occupation, with very strong elements of hunting remaining. Hodder sees this semi sedentary, semi developed farming still maintaining strong links with world of the wild animals and plants as causes for lack of any definite housing structure.

However, we must not forget that slights structures encountered on the sites, mainly with single occupational layer and spread over the large area, were probably the most appropriate in such still highly mobile system, still highly dependent on the wild. Also, relative poverity of material culture primarily identified in the pottery range, is often seen as the product of such circumstances. Nevertheless such strategy was after all successful and well adapted to survive and develop in subsequent stages of consolidation of farming during later fifth and further millennia.

Third group could include foragers adopting elements of agriculture, is visible at the sites as Odmut or Crvena Stijena in the Dinarids or Bug-Dniester foragers which had numerous influences of Koros and early LBK Neolithic culture especially during 5th millennium BC.

The earliest evidence in the Europe for the farming is encountered in the SE Europe. It also Yielded evidence for possible indigenous cultivation, and therefore is often regarded as very important research region, evidences from which are likel to change and clarify our picture about thedevelopment of the earliest agriculture.

Theories

The Fertile Crescent

The Fertile Crescent is a bow-shaped tract of land in Southwest Asia stretching from Jordan northwards to southern Turkey, then swinging southwards to the borders of Iraq and Iran. Along its route it incorporates parts of Israel, Lebanon and Syria.

The physical characteristics of this zone were formed millions of years ago when movement of the earth's crust forced the Arabian Peninsula toward the stable Iranian Plateau, compressing the land into pleats and folds.

The development of domestication, agriculture and sedentary communities in South-west Asia involved complex regional processes that crystallized over a large area of the Fertile Crescent.

At the end of the Pleistocene, the uplands of the Fertile Crescent supported large herds of wild sheep, goats, cattle and pigs and in many areas dense stands of wild wheat and barley. Also on lower elevations and in wetter regions were lakes and streams with abundant supplies of water-fowl, fish and other resources.

With the end of the Pleistocene 10,000 years ago, warmer, wetter weather patterns extended the ranges of early cereals into areas of South-west Asia where the first domesticated plants and agricultural communities were to appear. This phase of transition during the period 9,000 BC to 7,000 BC (from food-collecting to food-producing) has been termed "The Proto-Neolithic". It belongs to that stage of human development which witnessed the beginnings of village life and a shift away from subsistence based entirely upon hunting and gathering in favour of one over which man exercised more control : the era of "Incipient Cultivation".

Soon after 8,000 BC sedentary communities and domestic plants and animals began to appear in many areas of South-west Asia. These domesticates and allied agricultural economies were to prove both successful and adaptable to the extent that within centuries of their first appearance they had spread far outside the Fertile Crescent. By 7,000 BC farmers in Greek Thessaly were subsisting on cultivated emmerwheat and barley as well as domestic cattle and pigs.

Transition to Agriculture

Agriculture in Mesolithic Age

The Mesolithic has traditionally been seen as the period of hunter/ gatherers in Europe: small bands of nomadic wanderers living off the land and making use of natural resources rather than managing or altering their food sources to suit themselves.

Certainly there is no doubt that widespread exploitation of natural resources on a seasonal cycle was chiefly the manner in which these people subsisted. For example , at STAR CARR in the UK, we see a seasonal camp aimed perhaps at exploitation of Elk, Red Deer and other large ungulates, with considerable evidence of large scale hazelnut consumption.

We can see clearly across Europe that people were quite able to live on what they were provided with, especially during the early Mesolithic when the huge North Sea basin was dry and teeming with life. Some specialisation in each area is also obvious, for example seals, fish and shellfish were the dominant food source in some northern coastal areas ; there is even evidence for sea fishing in the dug out boats at MORTON (Scotland) and PESSE (Holland).

Plants played a vital role in most Mesolithic settlements, and can be seen from the fruit and nuts of the FRANCTHI CAVE (Greece), right up to the Baltic Sea where waterchestnuts and hazel are found in quite large quantities. Birds, waterfowl, small game and seaweed were also possible food sources too.

Evidence suggests however that as much as 80% of the nutrition of people came from larger game (Zvelebil), and this was certainly a huge part of most subsistance schemes, although the type of game hunted varied according to the region. Increasingly we are seeing evidence that some control ; often verging on the beginnings of agriculture, was exercised before the Neolithic. For example there have been suggestions of early selective breeding in Eastern europe, as at SOROKI (Ukraine), and there is evidence of clearance by fire at many British sites, eg IPING COMMON, as well as at some Aller¿d period Dutch sites. This has been suggested as an attempt to concentrate deer on fresh pasture, and could be described as incipient herding.

Large grass pollens have came to light at sites such as ICOANA (Romania), FLEA MOSS WOOD (UK), and WILLOW GARTH (UK), and it has been speculated that these could represent early attempts at utilisation of these plants, predating the introduction of better cereals from the Near East. Even some bone and antler implements could have been used as digging and planting tools, although this can not really be proven and there are other valid uses for many of them.

Many people have speculated that this attempt at control of their enviroment came about as a result of a change in their circumstances, such as the inundation of the North Sea which reduced available foraging land and concentrated a growing population around the shores of the new sea; this would necessitate a change in subsistence strategy.

However we could also speculate that contact with Neolithic groups caused some changes to occur, for many of the later sites are concurrent to, and even postdate in some areas the availablity phase of Neolithic technology, this being the case in Northern France and Central England especially. In Ireland we see the uptake of farming at CASHELKEELTY while all around foraging is dominant for nearly a Millenium; they had the ability but chose not to take it up, as is true of the ERTEBOLLE culture of Denmark too. This survived until almost 3000bc despite a millenium of contact with the Neolithic LINEARBANDKERAMIK culture only a few miles to the south, probably because foraging proved more profitable until the increased salinity of the Baltic Sea made it impractical.

In Sweden and isolated areas such as the Hebrides the Mesolithic way of life continued for a long time, in that foraging was the prime method of subsistence with only small support from agriculture. Even in other areas hunting and gathering continued as a support for agriculture though on a limited scale.

Central European transition to agriculture

Changes in the late Pleistocene and early Holocene climate have been cited as causing faunal and floral adaptations which lead to alterations in the subsistence activity of humans, namely an increased reliance on wild plant foods.

European agriculture, originating in the tell settlements of the Balkans and Greece in the seventh millennium B.C., was adopted or transported across most of the continent over the following two thousand years, later still before it appeared in Britain and Scandinavia. The initial spread into central Europe is thought to have followed the Danube to the region of Budapest in Hungary, hence Danubian.

The Early Agricultural Neolithic also witnessed the development of pottery technology; most of the modern divisions and sub-divisions within this period draw upon pottery styles (e.g. Linearbandkeramik culture) as indicators of cultural diversity or change within society. However, it is possible that farming economies developed independently of pottery, despite their apparent synonimity.

Settlement is an important feature of these early agriculturalist societies, often providing clear evidence of plant and animal domestication. The timber longhouse is an imposing symbol of the technological advances made during the early Neolithic. Artefacts are often found in or around these dwellings, notably figurines, decorated pottery, carved bone and stone tools such as sickles.

Three Initial Colonization Cultures have been identified as being responsible for the spread of agriculture over much of Europe. These gave rise to regional variations which took on distinct personalities and are traditions in their own right. The three cultures are:

— Starcevo-Körös

— Impressed Ware

— Linear Pottery (Linearbandkeramik or LBK)

The latter two are of most significance to the study of central Europe, the last in particular as it was the dominant culture throughout most of the region. Impressed or Cardium Impressed Ware as it is also known, is found on the Atlantic coast and in the south of France, and one inland cave, also in southern Spain. It has been suggested that it's derivation was across the western seaways from Portugal.

The Linearbandkeramik tradition is one of uniformity, in terms of artefacts, dwellings and settlement organisation and location. Plant remains from LBK sites point to small clearings in what must have been fairly developed forests, where early domesticates such as emmer and einkorn were grown in close proximity to the settlement.

Linear Pottery originating in an area west of the Danube appears to be responsible for the spread throughout Czechoslovakia, Poland, Germany, Holland and France to the Paris Basin. Danubians tilled the loess soils of lowland areas, perhaps practicing swidden or shifting agriculture as the soil of one region became exhausted. Some settlement evidence appears to support this hypothesis.

Neolithic of Mediterranean

The spread of agriculture throughout the Mediterranean area can be seen following a pattern which is very similar to that of cardial or linear ceramic ware.

Entering Greece by around 6000 BC from the Near East, it then proceeded to spread through out Italy, the South of France and finally becoming well established in Spain by the fifth and fourth millennia.

Early Greek sites are generally thought to have been aceramic but this absence of pottery does not detract from the ease at which the first main agricultural sites can be traced. Perhaps one of the most important, and earliest sites in Greece is the tell or mound village of Nea Nikomedia in Macedonia, this site, which dates from around 5800-5300 BC, gives a large variety of evidence for both plant and animal domesticates.

The main livestock at this site consisted of both goats and sheep with the main crop being emmer, other crops, for example, barley and einkorn were produced but on a smaller scale. The Franchti caves, situated in the Peloponnese, also offer a wide selection of good indications regarding the switch to agriculture and animal domestication. Around the same time that these new economical changes occurred in Nea Nikomedia, there is a large stock breeding society established in the Franchti caves, again specialising in sheep and goat breeding.

By around 5000-3000 BC a gradual change results in the quantity of emmer being produced to decrease to a level similar to the production of barley and einkorn resulting in a more balanced production of a range of crops, the land around the fertile Thessaly plains also becomes more exploited.

In Italy, many sites were established between 5000 and 3000 BC on the Tavoliere plains in the mid-south of the country, with the earliest Italian site dating from 5050 BC. Sites such as Monte Aquilone show signs of animal breeding with sheep, goats and cattle being well represented. In the north-central part of the country, just to the south of the Po river, there are signs indicating the mix of farming and hunting, but in the same area there is evidence of domesticated emmer and other crops along side domesticated animals by 4500 BC.

The cave site at Grotta del Santuario della Madonna also give rise to a new idea, one which comes from the quantity of obsidian found, from around the same time as the many sheep and goats at the site. Obsidian was unavailable on mainland Italy, but was available on some of the offshore islands, indicating the fact that the farmers of Grotta della Madonna were probably also fishing and communicating with the nearby smaller islands.

The earliest farming communities in the South of France were from around 4250 BC onwards, the most intensely farmed parts of this area were Provence and Languedoc, due to the fertile land in the immediate areas. Le Baratin, situated in the province of Languedoc was mainly an animal herding settlement, but there is evidence showing that the people of this site also underwent the exploitation of marine animals. Evidence of crop domestication is available from the sites of both areas, but one site imparticular, Baume de Fontbregoue, is host to a wider variety of crops than is usually seen, among the finds from the site is a large quantity of grain.

The area of Almeria in Spain also gives evidence for a possible new addition to the domestication of crops, this particular crop is olive vines. In the fourth millennium BC Spanish agriculture is based on wheat, barley, legumes, sheep and goats and by the third millennium BC there is evidence for the use of irrigation in some areas. It is known that there were sites in Spain as early as 5000-4500 BC, one such site is Coveta de L'or in Valencia, this cave site gives many signs of the usual, almost staple crops of this area, einkorn, emmer and barley.

Neolithic of the Atlantic and Northern Europe

During the later sixth Millenium BC we see the gradual spreading out of villages and homesteads from central , eastern and parts of southern Europe into the river valleys of the North.

This movement is associated with the linear pottery culture (or Linearbandkeramik - LBK) which is seen from about 4600 bc from Hungary and the Danube, and spreads rapidly along the rivers of the North German Plain and

Paris/ Rhine basins. Villages like ELSLOO in Holland, EITZUM in Germany and BRZESC-KUJAWSKI in Poland all appear with a remarkably similar material culture in a short time span, primarily in the Aisne/ Marne and North German areas.

These Cultures practised herding and had cereals like rye, emmer and barley available to them, although they displayed regional variation in their production according to conditions . Many also practised at least some foraging/ hunting behaviour as well.

During the 5th Millenium BC these concepts began to move onto the flood plains and eventually into some highland areas. However movement did not take place beyond a clearly delineated area, and Mesolithic communities along the coasts and in the ERTEBOLLE culture of Denmark continued a subsistence way of life; although there is evidence of communication between the two such as the introduction of pottery to Mesolithic cultures. Coastal frisia was unsuitable for agriculture due to its swamps and bogs, but other areas could perhaps have had Mesolithic economies sufficient to support the population, agriculture was therefore an unnattractive propostion.

In the same period we see a few sites such as ROUCADOUR, LA TRANCHE and DISSIGNAC appear on the Atlantic coast of France with evidence of domestic crops and animals in small proportions. LA TRANCHE could be contemporary with the LBK culture, but the spread of Cardial Impressed ware up this coast could indicate the influence of agriculture spread from the south of France instead. Following this we see the earliest spread of agriculture into southern England across the channel at an almost simultaneous time into Southern Ireland.

The C14 dates do not appear in either until the mid 4th Millenium, but by that time agriculture including cattle, sheep, pigs, goats, wheat and barley and other crops, is well established from Devon to Norfolk and across the centre of Scotland at sites like BALBRIDIE and NETHER KINNEIL (- this would indicate transmission of actual people across from Ireland into central Scotland via the North Channel). Central England and the North of Scotland continue to be mostly Mesolithic, and much of Ireland has an 800 year overlap of Mesolithic and Neolithic in many areas.

Gradually, we see the building of large structures such as causewayed enclosures, as at HEMBURY, BUDELSDORF in Germany, and SARUP in Denmark. These show the increase in complexity of these agricultural societies, that enables them to produce sufficient food to support workers on these projects. The enclosures have been suggested as cattle Kraals and grain centres so may or may not have some agricultural significance.

With this complexity comes considerable upland clearance as at PLOGUERNEAU and BARFIELD TARN, and in response to these more difficult soils increased use of ards and ploughs. These have been found at sites such as the SWEET TRACK and SEEBERG BURG-SCHISEE, and plough marks at SARNOWO and SOUTH STREET show a pattern of shifting cultivation in semi permanent fields.

While LBK sites show primarily a farming way of life we see in the third millenium sites that appear on the French coasts and in Brittany a mixed subsistance of domesticates and wild food. In the early agriculture that replaces the ERTEBOLLE culture in Denmark this mixture is also evident, and many coastal areas rely heavily on sealing and marine sources.

In Sweden and the Baltic there is a brief introduction of this partial agriculture by the TRB culture around 3000bc, but a decline in climate saw this replaced by hunter/ gatherers until c.2300 when another group of settlers appear. Northern Scandanavia and parts of Northern Scotland do not see any agriculture for a considerable time after this.

Contemporary to these events we see in the the UK and parts of continental Europe an abandoment of much upland arable farming around 3200-3000 bc, due to the development of heathland.In some areas this leads to increased pastural farming, but in others we see a regrowth of secondary forest and grasslands, accompanied by movement on intensive agriculture back into the valleys.

Domestication of Animals

Animals were of vital importance to prehistoric farmers. They provide assistance with farm work, clothing, protection, as well as food. Land management became easier and tasks quicker. The disadvantages were primarily to do with feeding the animals. The carnivores often hunted the same species as man, whilst the herbivores had to have food grown for them, this food would of taken up space that human crops were to be grown upon.

Pig

Pigs were a native species to Europe and the Near East. and it is thought to of descended from the wild boar. The wild pigs were larger than the domesticated species and the reasons for their size reduction is not known. Barker thinks that it is not due to selection by man. Pigs had several advantageous features. The entire animal is edible and their reproduction rate is extremely high. Coupling this with a pig's ability to eat practically anything, resulted in a adaptable animal that was easy to rear and look after.

Cattle

Wild species of cattle were much larger than domesticated. This is probably due to natural selection of larger boned wild species, better adapted to survive in harsh conditions. Humans influence altered this trend. Poor quality winter feed, restricted grazing lands and selective breeding all brought about a clear dichotomy between wild and domestic species. Cattle were a prised species, able to provide: butter, milk, cream, and cheese whilst living. Their phenomenal power was utilised to plough fields and fertilisation of the land was another added bonus. Once dead cattle provided large quantities of meat and their hides could be use for clothing. Thus cattle were a vital resource to prehistoric farmers.

Sheep and Goats

Domesticated sheep and goats have a cloudy origin. They were first herded in the Near East. The advantages of sheep and goats is their hardiness. It is possible to keep the animals on very poor land with little grazing potential. They are able to survive out door at all times of the year and give birth naturally, without human assistance - although the survival rate was not high. The prehistoric sheep were smaller than todays and were of poorer quality. Sheep provided wool for clothing, milk and cheese and also meat. Their hides could also be used once the animal had died. Goats offered the same advantages as sheep. Their hides were especially valued due to their hard wearing properties.

Other animals

Other animals that were used in prehistoric farming included horses (Equus caballus), donkeys (Equus asinus), dogs (Canis familiaris) and red deer (Cervus elaphus). Horses were smaller than todays and were used primarily for transport i.e. pulling carts and riding. They were occasionally eaten, but due to their position as a prestigious animal this was rare.

Donkeys were used mainly as pack animals in long distance transport. They were occasionally used for light ploughing. Dogs were tamed considerably before farming was introduced. They originate from wolves and were larger than modern species. They were kept for their abilities to herd, guard and hunt. Red deer were a vital food resource for people in prehistoric times. They varied greatly in size and could live in deciduous forests or open moor land. The ability to herd them was necessary as hunting often took too long and required a specialised knowledge of local landscapes. The red deer were killed for their meat and also their antlers, which could be used in the manufacture of tools and weapons.

Domestication of Plants

The domestication of wheat and barley - economically the most important plants in South-west Asia - involved a complex series of changes in at least four native wild grasses, including:

— Wild barley - Hordeum spontaneum

— Wild einkorn - Triticum boeticum

— Wild emmerwheat - Triticum dicoccoides

— Goat face grass - Aegilops squarrosa.

Domestication of these plants, as elsewhere, was a long process in which certain kinds of mutations were perpetrated, mainly through changing patterns of exploitation.

For example the domestication of wheat, one of the world's most important crops, involved both human manipulation and natural hybridization between related genera. In this case human intervention appears to have been aimed at producing free-threshing, non-shattering varieties; and by mixing genetic material from various species, early farmers produced forms of wheat that could adapt to diverse habitats and conditions.

Wheats are the universal cereals of Old World agriculture. Together with barley, they constituted the principal grain stock that founded Neolithic agriculture and the main element responsible for its successful spread. Since this early start, wheats have retained their crucial role in Old World food production and have given rise to numerous advanced types.

Today wheat ranks first in the world's grain production and accounts for more than 20% of the total food calories consumed by humans. It is now extensively grown throughout the temperate, Mediterranean type, and sub-tropical parts of both hemispheres.

5

Managing Natural Resources

Natural resources are the threads of our social fabric. The interests affected by natural resource management are very diverse. They include the forest industry; indigenous peoples; policy makers in municipal, provincial and federal governments; landowners; consumers; affected citizens and affecting citizens, and so on.

Among the rainfed lowland rice areas in Asia, the Chhotanagpur Plateau in India constitutes an area of particularly low agricultural productivity and a high incidence and severity of poverty. The incidence of poverty among rural households in the area is estimated to be among the highest in Asia. Most rural households practice subsistence farming under adverse and risky environmental conditions. The natural resource base can be characterized as poorly suited to agriculture because of climatic, water resource, and soil conditions. Because of several decades of nonsustainable land-use practices and highly erosive monsoon rains, deforestation and soil erosion are proceeding rapidly in many parts of the plateau.

The families that live and work in the area often depend on nonagricultural income-generating activities to sustain the household-especially during the postmonsoon season. The majority of these nonfarming activities involve low-productivity and low-paying work. Trends with respect to the state of natural resources combined with the paucity of options for high-return nonagricultural employment make the prognosis for the future of the area, and of the families that live there, worrying.

LOCATION AND HISTORY

Physiographically, the study area is located at the eastern fringe of the Chhotanagpur Plateau, which rises in elevation south and west of the eastern Gangetic plain and the Ganges delta. Before the British colonial era, this area was largely unsettled and was mostly covered with dense forest. Available oral history and limited documentation suggest that the small local population subsisted from the extraction of local forest products and carried out very little agricultural activity, focusing on cultivating local rice, pulses, millets, and vegetables. Under British rule, the colonial administrators found the area to be an important source of timber for use in constructing national railways, leading to extensive deforestation. Settlers-particularly families from marginalized tribes-were permitted to move into the area to farm on the cleared lands. Although records are scarce, it seems that most immigrants into the area came from other parts of eastern India.

With independence, state initiatives established a mining industry in Giridih District, which was rich in mineral resources (particularly mica and coal), as part of a broader development strategy highlighting industrial development. The mining industry flourished for about 30 years, attracting more immigrants, until the demand for mined mica declined because of the invention of a less costly artificial substitute, while other mineral resources became depleted. With the decline of the mining industry, many of the immigrants turned to agriculture for their livelihoods.

On the other hand, Purulia District of West Bengal, serving as a buffer zone between the Damodar plains and Bihar plateau, came under British rule in the last decade of the19th century, with local zamindars governing natural resources, including forest, and peoples' livelihood. Except for a few coal areas, no important industries flourished. The main crops were more or less the same as those of Giridih District and tribals formed a significant part of the population. Postindependent reforms, particularly panchayat activities and land reforms, with an increase in small landowners, had major impacts in Purulia in the 1990s. For example, high-yielding varieties (HYVs) along with inputs such as fertilizer and pesticides were distributed among farmers through the panchayat under the scheme called "Manikit," which might explain the higher rate of HYV adoption in Purulia than in Giridih. However, the topographical features and the predominance of rainfed cultivation are common between the two districts.

Natural Resource Degradation and Unfavorable Environmental Conditions

Climatically, the study area is subtropical and subhumid, with hot wet summers and cool dry winters. During most of the year, temperatures in the study area do not constrain crop growth, although winter temperatures do fall below 10 C in December and January. Low temperatures can affect establishment of boro (winter season) rice, which is transplanted in December-January and harvested in April-

May. Rainfall and water availability are more serious climatic factors constraining agriculture in the study area. Annual rainfall generally decreases from the northwest to the southeast of the plateau. The mean annual rainfall in Purulia District is 1,334 mm, decreasing to 1,225 mm in Giridih District. While these annual amounts would generally be considered adequate for agriculture, the high seasonality of rainfall confines the cropping period to only part of the year. This rainfall seasonality is influenced by the southwest monsoon. There are, on average, about 80 rainy days (with daily rainfall exceeding 2.5 mm) in a year in Giridih and 83 in Purulia, but this can vary yearly, from 62 to 104 days. The monsoon rains can be very intensive and erosive, particularly in areas on higher ground bare of vegetation cover. Available moisture over the entire monsoon period determines the time window of opportunity for the various cropping systems practiced by farmers in the study area.

Traditionally, farmers in the study area use the appearance of the Rohan star in SMW20-21(mid-May) as an expectation of premonsoon precipitation for carrying out nursery planting and land preparation for rice. The average premonsoon precipitation (or Rohan rain) is 35.2 mm in Purulia and 26.9 mm in Giridih. Similarly, the availability of postmonsoon precipitation (Hatia rain) during the period of appearance of the Hatia star in the last week of September to the first two weeks of October (SMW40-41) indicates the possibility of sowing winter crops and also using residual moisture to cultivate winter crops on medium land. The field experiments indicate that second cropping is possible on medium land with soil residual moisture provided that there is Hatia rain to germinate and establish the crop.

Variation is considerable in the onset and withdrawal of the southwest monsoon. The onset can occur as early as mid-May (SMW20) or as late as mid-July (SMW28). The withdrawal of the monsoon can vary from late August (SMW35) to late October (SWM43). This large variation in both the onset and withdrawal of monsoon causes much uncertainty for farmers in implementing their cropping calendar in the study area. The total rainfall for 1996 (1,299 mm) was just 1.8% higher than the 27-year average (1,276 mm), while the 1998 rainfall was 1,672 mm, or 31% more. Furthermore, the seasonal rainfall pattern for these two years is vastly different from each other and from the average 27-year pattern. In 1996, the June rainfall was 36% higher than normal, but declined drastically in July, to only 35% of the 27-year average.

There was also an early withdrawal of the monsoon; the October rainfall in 1996 was 60% below the 27-year average (83 mm) and no rainfall was recorded for November. In 1998, the off-monsoon months of January to April were distinctly wetter and the onset of the monsoon was delayed-the June rainfall was 70% below the 27-year average. However, the monsoon lasted longer and the September and October rainfalls were 56% and 174% higher than normal. A delay in the onset of the monsoon means that farmers who take advantage of the Rohan rain to start their crops would encounter early-season drought. An early withdrawal and/or lack of Hatia rain would affect the ability to grow winter crops in a multiple cropping system. Hatia rain (over a 2-week period in late September to early October) can vary from 0 to 334 mm, and may fail completely in one out of four years. On average, only about 10% of the total annual rainfall occurs outside the monsoon (pre- and post-) period. Although this constitutes a small amount of rainfall, it is nevertheless important in giving reasonable yields for winter crops such as barley and oilseeds.

On the whole, differences in rainfall characteristics between Purulia and Giridih are only slight, with Purulia experiencing marginally higher rainfall and more rainy days, and slightly higher frequency of early monsoon onset (5 out of 20 years compared with 2 out of 20 years for Giridih). These minor differences across space are overshadowed by the considerably greater interannual variations at any location in the plateau.

Surface water and groundwater resources

During the dry season, many of the smaller streams/ rivulets in the study area typically dry out; as a result, surface water is not readily available. The main sources of water for agricultural use are dug wells, supplemented with water stored in reservoirs and ponds. The study area is poorly endowed with underground water resources because of its crystalline (granitic) bedrock and insufficient recharge from rainfall during the monsoon season because of excessive surface runoff. The water-table depth varies from shallower than 3 m to more than 12 m during the summer months, and many of the shallow wells retain scanty water or dry up completely. Hence, irrigation potential during the dry season is relatively low.

Influence of topography and soils on land use

Topographically, the study area is undulating, with elevation ranging from 250 to 500 m. The undulating topography and highly dissected landscape give rise to short-range variations in terrain and soil and water conditions, which influence the kinds of crops that can be grown, the time windows for cropping, and the possible cropping systems in different parts of the toposequence. Topographically, three main landscape types are identified:

— The upper terraces, or uplands, are associated with light-textured soils.

— The middle terraces, or medium land, have slightly more heavily textured soils (e.g., loamy sands and sandy loams).

— The lower terraces, or lowlands, have even more heavily textured soils (e.g., sandy clay loam).

Soil texture influences soil moisture storage capacity, which, together with water-table depth, largely determines the period of moisture availability for cropping at these landscape positions. Water availability outside the monsoon season (including pre- and postmonsoon periods) is particularly crucial for cultivation in the upland and medium land, whereas, in the lowlands, the accumulation of excessive surface runoff coupled with a high water table causes flooding, which could keep the land submerged after the monsoon period until as late as February. Farmers further distinguish different levels among the upland, medium land, and lowland, planting different rice varieties according to their perception of soil moisture gradient across the toposequence.

The uplands, locally known as tanr, are subjected to various intensities of use.

1. The gravelly uplands are generally not suitable for cultivation because of the harsh soil and water conditions. Where the natural forests are degraded, these uncultivated areas undergo severe erosion, causing further land degradation and increasing the extent of uncultivable area.

2. In the cultivable uplands, cropping of gora rice (short duration, 85-90 days, drought-tolerant, low-yielding

upland rice) and traditional minor millets (finger millet, kodo, gundli) is done solely or in association with other crops as mixed crops during the monsoon season. Across all the sample villages, the most common cropping pattern for uplands was rice (kharif or monsoon-season)-fallow (rabi or winter-season), which was practiced on 41% of the total upland plots (in terms of the number of plots). Seventy-four percent of the upland plots are left fallow during the winter season.

3. The uplands close to the homesteads, called barhi, are supplied with water from dug wells. The barhi land is intensively used to cultivate vegetables, high-yielding varieties of potato, and wheat and rice, and for rice seedbeds. The relatively higher level of inputs (including labor) in these barhi lands compensates for the low natural fertility of the soils and enables these areas to support crop diversification and intensification. In contrast with other land types, the rice-fallow cropping pattern is found on only 4% of the barhi plots, and only 36% of the barhi plots are left fallow during winter. Cropping patterns practiced on barhi lands are quite diverse, but the more commonly observed ones during the kharif-rabi seasons on barhi plots were fallow-potato (16%), maize-fallow (15%), ragi (finger millet-Eleusine coracana)-fallow (7%), maize-potato (7%), rice-fallow (4%), ragi-potato (4%), and fallow-wheat (4%).

4. The lower portions of the uplands (mid-uplands) and the medium lands are distinguished by the rice varieties that are grown, as influenced by soil moisture availability. In the mid-uplands (or baad), medium-duration rice varieties are grown, whereas, lower down

the toposequence in the medium land where soil moisture is available for a longer period, long-duration rice varieties are grown. These medium lands are known as *ajan* in Giridih and *kanali* in Purulia. Winter crops such as rapeseed, linseed, barley, lentil, and gram can be cultivated successfully following the monsoon-season rice crop, using residual soil moisture, particularly in the miduplands. In the medium lands, the most common cropping pattern is still rice-fallow, which covers 87% of the total medium land plots.

5. Although the lowlands are the most fertile, excessive moisture and poor drainage limit cropping to only rice, and constrain growing winter crops after the main rice season. Presently, most farmers grow traditional longduration rice varieties with low inputs of manure. The rice-fallow cropping pattern is practiced on as much as 91% of lowland plots.

In general, the plateau soils are low in organic matter as well as available phosphorus, and have medium levels of potassium (150 to 300 kg ha). The soil is slightly acidic in some areas. The soil fertility characteristics also vary with landscape position and with intensity of cultivation. There is a marked difference in the soil fertility level of the barhi and tanr in the uplands because of the higher inputs applied to the barhi lands, which are the most intensively cultivated. Apart from the barhi land, there is a systematic trend of increasing soil fertility from the uplands to the lowlands.

Because of the risky environment and the relative difficulty in gaining access to markets, food security is a primary concern of households in the study area. As a result, large amounts of the resources of farm households (especially family labor) are typically applied to rice production, particularly in the rainy season, for their own consumption. Various surveys show that, except for the case of barhi lands, most of the other plots are left fallow during the nonmonsoon season. However, this does not mean that the household labor force is left idle; many members are engaged in a wide variety of nonagricultural income-earning activities during the agricultural off-season, as will be described in the section on "High incidence of poverty."

Agricultural production

Low productivity and limited use of modern technologies Agriculture in the sample villages is largely ricebased, with a very strong subsistence orientation. The average yield of paddy across all land types, seasons, and villages during the survey years was 2.8 t ha. This yield level, however, appears to compare favorably with eastern Indian standards based on aggregate data. The average rice yield during 1990-97 across seven eastern Indian states was 2.0 t ha. The average paddy yield within the eastern Indian plateau (covering the plateau area of Bihar, Orissa, West Bengal, and eastern Madhya Pradesh states) during 1995-97 was 1.6 t ha.

The average rice yield also observed across surveyed villages ranged from 2.3 t ha(Kumardi village in Purulia District) to 3.2 t ha(Fateha village in Giridih District). Because of the higher intensity of input use as well as differences in soil fertility and production systems in lowland plots and land on the higher levels of the toposequence, rice yields are highest in plots on the lower terraces or lowlands. Average rice yields in upland, barhi land, mid-upland, medium land, and lowland were 2.1, 2.5, 2.5, 3.0, and 3.3 t ha, respectively. The average yield of high-yielding rice varieties is higher than that of traditional varieties by 20% to 40% depending on land type and this yield differential tends to be larger on upper terraces than on lower terraces. Pair-wise tests of the difference in means (t test) further reveal that per hectare cash expenses are significantly lower on upland than on medium land and lowland, and cash expenses on mid-upland are lower than those on lowland. The average per hectare value of rice produced is significantly lower on upland and mid-upland than on other land types. The net rice income (before subtracting the shadow value of the cost of family labor) is higher on lowland and medium land and lowest on upland. With family labor costs (valued at observed market wage rates) included, however, the average net income (after subtracting both cash and family labor) turns negative for the upland, barhi, and mid-upland, while remaining positive for medium land and lowland. This is a result of the high shadow value of the labor cost that, in turn, is due to the high labor intensity of the production system, especially on barhi plots.

The use of modern technology in agricultural production was relatively limited among the sampled households. The rate of adoption of improved crop varieties (HYVs), measured by the share of the households adopting HYVs, ranged from 3% (among wheat-cultivating households) for wheat to 51% for rice, on average, across all the sampled villages. This compared unfavorably to the estimated 70% HYV adoption rate for rice in eastern India as a whole.

Furthermore, based on the plotlevel data, only 21% of the rice planted area during the kharif season was planted with HYVs. The HYV adoption rate, however, differs somewhat across land types; the HYV adoption rate is roughly twice as high on medium land and lowland as on upper terraces.

The main reasons provided by interviewed households for why they did not use HYVs were the higher risk and higher seed cost involved. The adoption of other rice production practices associated with more modern cultivation practices, such as the practice of intercropping, use of purchased chemical inputs (e.g., fertilizer, pesticides, or herbicides), and use of irrigation, further suggests the traditional nature of agriculture carried out by the surveyed farms. Intercropping was reportedly practiced by only 24% of sampled households, but the adoption rates contrasted sharply across surveyed households in Giridih (42%) and Purulia (7%) districts.

In response to a question about why intercropping was not practiced, the predominant reason was "lack of know-how." Application of pesticides in rice crops was relatively low (42%)-with the main reasons for nonuse (as provided by survey respondents) being lack of funds or lack of know-how.Less than onethird (31%) of the surveyed farms reported applying urea. Less than a third (roughly 30%) of the households reported using irrigation and the main reason for not using irrigation is reportedly "lack of resources" (i.e., money for pumping equipment, inaccessibility of water).

Subsistence orientation

Food security appears to be the prime objective of the surveyed farms, resulting in their strong orientation toward subsistence in production. Anecdotal evidence and oral histories suggest past vulnerability of the sampled households to years of hunger and starvation, which would explain the fixation on satisfying food security. While 90% of the households are reportedly engaged in rice production, for example, only 21% reported selling of rice during the year covered by the survey. Similarly, 22% of the households produce potato, but only 2% reported selling potato, and 21% produce but none reported selling maize.

The subsistence orientation of surveyed farm households is further demonstrated by the fact that, across all crops, only 23% of the surveyed households reported the sale of any farm output. The sale of farm outputs provided only 3% of the total household gross income, on average, whereas the imputed value of home-consumed farm outputs equaled 34% of the total household gross income.

The surveyed farms' activities in agricultural output markets predominantly involved transactions in the rice market. Among the total number of reported incidences of crop sales among surveyed farms, 49% involved rice, 12% potato, 11% maize, 9% millet, 7% horsegram, and 7% wheat. The concentration of product sales on nonperishable grains indicates that transportation costs (i.e., poor transportation infrastructure makes it costly and time consuming to transport goods to market) may constrain commercial agricultural activities in the study area, and the focus on rice suggests that commercial markets for crops aside from rice are quite limited.

Subsistence orientation also dominates the surveyed farms' animal husbandry activities. The imputed value of home consumption of livestock and poultry (including both "main products" such as milk and eggs and "by-products" such as animal dung) averaged Rs. 3,565 (about US$90 at the 40 Rs. = $1 exchange rate that prevailed in 1997), while the average income from livestock sales was only Rs. 570 (roughly US$14) across all households. The sale of livestock/poultry was reported by only 19% of the households and income from such sales made up only 2% of the total gross income on average. The total imputed value of home consumption of livestock and poultry equaled 13% of the total gross income. The participation of surveyed households in markets for agricultural inputs was also low. Farmers generally appear reluctant to use purchased inputs or to make investments because of the high risk of crop failure-which threatens loss of costs of purchased inputs and investments-, because of the high transactions costs (i.e., inaccessibility to markets makes it costly to purchase inputs in markets) and because of the low adoption of HYVs. The rates of application of purchased chemical inputs were reviewed above.

The share of the households reporting hiring of paid labor across the principal tasks required for rice cultivation was 14% (for weeding), 36% (for threshing), and 40% (for transplanting). The rental of capital equipment or draft animals among surveyed farms was also infrequent; the shares of households reporting having rented or leased land, draft animals, or various machinery were 0%, 13%, and 6%, respectively.

High incidence of poverty

Income and wealth

A defining characteristic of the households in the study area is their predominantly low standard of living and high incidence of poverty. According to official poverty estimates of the Indian government, the incidence of poverty in the study area was among the highest in the country. Statewide headcount poverty ratios in Bihar and West Bengal were the second and third highest in 1987-88, first and fifth highest in 1993-94, and second and fifth highest in 1999, respectively. Among all 16 sampled villages, the estimated average annual household gross income was Rs. 27,269, while the average per capita annual gross income was only Rs. 4,018 (about $100).

The village average per capita gross income among surveyed households ranged from a low of Rs. 3,003 (in Naitanr village in Giridih District) to a high of Rs. 6,474 (in Charak Patla village, also in Giridih District). The planning commission's official poverty line for 1999 and the estimated per capita gross income indicated that 60% of the sampled households were poor. Across the surveyed villages, the estimated headcount poverty ratio ranged widely, but all had a substantial portion of their population considered poor. Poverty incidence was lowest in Charak Patla village (24%) and highest in Patharkata village (82%). Average per capita gross income is the same for Giridih (Rs. 4,026) and Purulia (Rs. 4,010); however, because of the higher cost of living-thus, higher local rural poverty line-in West Bengal, the estimated headcount poverty ratio is higher in Purulia (63%) than in Giridih (57%).

Another basis upon which to draw conclusions about the standard of living enjoyed by surveyed households is to consider their asset holdings. Survey results suggest that asset holdings among sampled households were quite limited. For example, about one quarter of the sampled households reported owning no household appliance. Among major household appliances, only 23% of the households reported owning a radio and only 5% reported owning a TV set. Just under a third (32%) of the sampled households reported owning animal-powered transportation, not a single household reported owning an automobile, and only 4% reported owning a motorcycle. Most households (78%), however, did report owning human-powered transportation (such as bicycles). For complete details on household ownership of appliances and equipment. Ownership of agricultural implements and machinery was similarly limited, according to the survey results. The total average value of agricultural implements across all households surveyed was Rs. 3,035 (equivalent to about $76). Village-level means ranged from Rs. 566 to Rs. 18,528-mainly because of variations in the ownership of tractors, rice mills, and water pumps, which are quite rare in any case.

Ownership of mechanical threshers was reported in only five villages, and water pumps were found in only seven of the 16 villages. Tractor ownership was similarly rare, with tractors being reported in only three villages. Bullock carts were found among surveyed households in all villages and rice mill ownership was reported in only one village. Ownership of smaller hand tools and other common agricultural implements was found in nearly all the households surveyed. Ninety-six percent of the sample households reported owning either hand tools or animal-driven equipment.

Survey results suggest that livestock are held mainly for the draft power they can provide (i.e., for plowing or for pulling a cart) and as a form of savings, but they are rarely sold in the market. The mean value of total livestock holdings was Rs. 7,926 across all households surveyed. Nearly two-thirds (64%) of the sampled households reported bullock ownership, and on average they owned two bullocks per household. The reported ownership rate of other smaller livestock or poultry was lower but still significant. Surveyed households reported owning goats, chickens, buffalos, cows, and calves at rates of 46%, 16%, 15%, 39%, and 11%, respectively.

Other indicators of household welfare

Apart from the level of household income and asset

holdings, the living standard can also be measured by household satisfaction of basic needs. In terms of access to education, results from the 1991 census show that the incidence of illiteracy in the area was high: 65% in Purulia and 80% in Giridih. The average years of schooling among the household heads was 3.6 years, with Purulia again displaying better levels of educational attainment than Giridih (3.9 years in Purulia versus 3.4 in Giridih).

Access to other basic public services also appears low, according to survey results. Less than 12% of the sampled households had access to electricity. In seven of the 16 surveyed villages, none of the sampled households had access to electricity. Among the nine other villages, the proportion of households with access to electricity ranged from 3% to 33%. The source of drinking water for the majority of households is dug wells (54%), but other sources include tube wells (20%), "tanks" (ponds) (9%), rivers (8%), temporary dug wells (5%), and the tap (piped-in water) (2%).

On average, households reported that a walk of 11.4 minutes is required to reach their nearest source of water. This represents a significant labor burden on female members of the household, who are principally responsible for fetching water for household needs. These travel times vary markedly across surveyed villages, with a low of 4.5 minutes in Parsatanr village, where most households have access to dug wells, and a high of 27 minutes in Palkia, where the only water source for most of the households is a neighboring river. The inaccessibility of most of the surveyed villages, as mentioned earlier, also suggests that access to other public services (e.g., health care, legal institutions, social welfare) is difficult and time-consuming.

Sources of household income

Role of nonagricultural income

Diversification of income sources is one of the most prevalent household strategies for coping with risk and vulnerability in rural areas with less favorable economic and agroclimatic conditions. Looking at the primary occupation of workingage household members reported in the Purulia survey, the most commonly observed occupation is "unpaid family labor on family farm," which was reported by 50% of the total individuals across the surveyed households that reported having a principal occupation.

The next most common principal occupation of surveyed individuals was miner (20%), followed by casual day laborer (10%) and paid full-time or parttime agricultural laborer (4%). For secondary occupations, the most frequently reported occupations were, again, unpaid family labor on family farm (45%), casual day laborer (19%), and paid full-time or parttime agricultural laborer (13%).

These results suggest that the economic activities of the sampled households are largely dominated by unpaid family labor on family farms and in agriculturally related jobs, but also show that there is significant diversity in the occupational activities of surveyed households. This diversity is even clearer when we can examine this issue from the perspective of the share of household income derived from different types of employment. Results make clear that the share of the households engaged in formal labor markets is relatively low-a little more than one-third of the male workers (35%) and onequarter of the female workers (28%) reported being employed in wage-paying jobs according to the survey.

Caste system

Ethnically, the Chhotanagpur Plateau lies in what is commonly referred to as the tribal belt of eastern India because of the high proportion of individuals from scheduled tribes (ST) and scheduled castes (SC) in the population. Under Indian law, citizens from ST or SC are eligible for targeted public assistance and political representation in state and federal legislatures of ST and SC is guaranteed. These measures are intended to remedy past discrimination against ST and SC and to facilitate full participation of ST/SC members in the economy and polity.

Across the sampled villages, households of ST or SC background constitute an average of one-third of all households surveyed. The share of ST families was higher in Purulia than in Giridih, whereas the share of SC families was higher in Giridih than in Purulia. In general, survey results suggest that villages in Giridih tend to be more segregated (or to have more homogeneous populations defined in terms of their caste background), whereas villages in Purulia had more heterogeneous populations in terms of caste composition. There was wide variation in the share of ST/ SC families across the sampled villages. Categorizing surveyed households into four caste groups: ST, SC, other backward castes (OBC), and other castes (OC)-which includes all higher castes, in two villages (Charak Patla and Palkia), the entire population of households surveyed came from ST families. Both of these villages are in Giridih District. Two other villages (Fateha and Parsatanr-both in Giridih) were exclusively inhabited by households of SC and OBC backgrounds. At the other extreme, other surveyed villages displayed a high degree of heterogeneity in the caste background of the inhabitant households. Three villages had no single dominant caste category among the households interviewed.

It is also noteworthy that tolas dominated by ST families tended to occupy more marginal lands, with many farm plots located on unproductive marginal upland. Other villages represent intermediate cases involving villages whose inhabitants came predominantly from OC households (Fulchi and Naitanr villages in Giridih) or a mix of SC or OBC dominant villages with significant representation of other caste groups. The relationships between a household's caste affiliation or the caste structure of a village and various economic outcomes appear complex. While one might expect that households from higher castes would have higher income or enjoy a higher standard of living, only some indicators present in the survey data suggest that this is the case, while other indicators suggest that less advantaged castes enjoy better economic outcomes.

As expected for example, none of the large farm households belonged to scheduled castes, and the share of households from the scheduled caste background tended to be higher among the landless and marginal farmers than among the small or medium farmers. Within villages, some association between landholding size and caste, that is, that higher castes would control larger landholdings. This tendency was observed to some extent in Giridih but not in Purulia.

Overall, the correlation between landholding size class and caste was not particularly strong in the survey data. Considering the overall size of landholdings (regardless of land type), the majority of sampled households belonged to the three smaller landholding classes (i.e., landless, marginal, or small), regardless of their caste affiliation. Households with an OC background were not significantly more likely than families from lower castes to own larger land areas. For example, in Purulia District, the average landholding size among OC households was about 0.8 ha, which was larger than the average landholdings of ST or SC households, but was smaller than the average landholdings of OBC households in the district. In Giridih, the average landholding size among OC households was larger than the average landholdings of SC households, but was close in size to the average landholdings of ST or OBC caste households.

When land is considered in terms of its agricultural potential, however, a slightly different picture emerges. Survey results show a systematic tendency of greater average size of lowland held by OC households in Giridih District (roughly 0.4 ha) vis-à vis the amount of lowland owned by households from the disadvantaged castes. This tendency was not observed in Purulia, so this supports the conventional wisdom that disadvantaged castes tend to suffer greater marginalization in the more traditional Giridih District (formerly Bihar state).

Size of landholdings and economic outcomes

A positive correlation existed between per capita or total gross income and landholding size among the surveyed farms. Similarly, poverty incidence was higher among households with smaller landholdings, but there was no difference in the level of poverty incidence between landless households and marginal farm households. Survey results showed a positive correlation between landholding size and the total value of livestock held. A positive correlation also existed between landholding size and ownership of some appliances (e.g., radio, TV, electric fan) or a motorcycle. The share of households reporting owning no appliances was from 20% to 30% across all landholding classes. Ownership of basic farm equipment (such as the desi plow) displayed relatively little variation across farm size, but households with larger landholdings were more likely to own power-driven equipment such as water pumps.

The agricultural practices used by surveyed families were observed to vary systematically depending on landholding size. The adoption of intercropping, use of irrigation in rice crops, and use of "modern" inputs or practices (e.g., application of chemical fertilizer or insecticides/pesticides on rice) all tended to be positively correlated with landholding size. Nevertheless, there was little systematic relationship between landholding size and many other agricultural practices. In particular, no relationship was found between landholding size and the adoption rate of HYVs.

These observations suggest that landholding size (unadjusted for land quality) was a poor measure of the agricultural potential of households' farms because of the disparate productive characteristics of different land types.

The impact of land reform and the panchayat system

Although Giridih and Purulia districts share similar agroecological characteristics, the political systems and governing institutions in the two states differ markedly. West Bengal State came under the Left Front government in 1976, which subsequently implemented a major land reform program. The land reform placed an area ceiling on legally permissible landholdings, redistributed land to the landless, and conferred secure tenancy rights. This was accompanied by the implementation of a system for democratic local governance, the panchayat, at the village level. Bihar State, in contrast, underwent no serious land reform and the panchayat system has not been vigorously implemented there.

In Bihar, large landholders tend to exercise considerable influence over economic and political institutions down to the village level. Given the contrasting histories and resulting differences in institutional environments, expect some systematic differences in the socioeconomic outcomes (e.g., level of poverty, agricultural practices, etc.) between the two districts. The caste structures of the villages demonstrate more villages with mixed caste compositions in Purulia, while villages are more segregated (including single-caste villages) in Giridih. This difference could trace its origins to the political developments mentioned in the prior paragraph. Historically, Giridih was settled earlier, but population density is higher in Purulia.

For observed agricultural practices among the surveyed farms, several differences are seen. One major difference is that the intensively cultivated upland homestead plots (barhi land) that rely on a heavy use of labor and irrigation are widely observed in Giridih, but relatively few such plots tend to be found in Purulia. In rice production, HYV adoption, together with the use of modern chemical inputs (fertilizer and insecticides/pesticides), is somewhat more widespread in Purulia than in Giridih. In Purulia, 66% of the surveyed farm households reported using HYVs compared with only 39% of the surveyed households in Giridih. The main reason given for nonadoption in Giridih was the high cost of seed, while the main reason for nonadoption in Purulia was the higher risk associated with HYV cultivation. The rate of adoption of fertilizer in Giridih was 55% and in Purulia it was 66%, whereas insecticides/pesticides were adopted by 37% and 47% of the households surveyed in Giridih and Purulia, respectively.

The establishment of rice crops using transplanting was slightly more prevalent in Giridih (81%) than in Purulia

(76%), whereas direct seeding was more common in Purulia (14%) than in Giridih (4%) (the remaining households adopted "behind the plow" seeding). The surveyed farms in the two districts irrigated their rice crops at similar rates. Mixed cropping/intercropping was more widespread in Giridih than in Purulia. While it is possible that differences in the availability of agricultural extension services or competitive input markets are associated with political and economic differences between the two districts, these different practices could also be due to differences in soil types and common landscape positions of farm plots, which influence soil moisture and weed conditions, across the two districts.

The "capital poor" nature of the farm households, on the other hand, was similar across the surveyed households in the two districts. Most of the households in both districts reported owning basic hand tools (88% in Giridih, 93% in Purulia), while a small minority reported owning diesel water pumps (3% in both districts). Despite some observed differences in agricultural practices, average rice yields across the two districts were roughly equal (2,824 kg hain Giridih and 2,663 kg hain Purulia). Ownership of animals appears to be somewhat more common in Giridih than in Purulia, and this could be because animal rental markets (especially rental of bullocks) appeared to be more active in Purulia. Also, in Purulia, year-round holding of bullocks is constrained by the lack of grazing land due to the diminished availability of common grazing land in the district. On the whole, rental markets for livestock were thin according to the survey results.

The average values of livestock holdings were higher in Giridih (Rs. 9,076) than in Purulia (Rs. 6,822), but the total income from livestock sales was higher in Purulia (Rs. 967) than in Giridih (Rs. 159). In terms of credit markets, households in Purulia tended to rely more on informal sources (neighbors and money lenders) than households in Giridih (where credit through commercial banks, the Gramin Bank, and state sources was more common). This runs counter to the a priori expectation that economic transactions would be more formalized in Purulia than in Giridih because of a higher level of institutional development in Purulia resulting from the more advanced development of the panchayat system in that district.

Labor market participation among male workers was somewhat higher in Giridih than in Purulia, but was about the same between the two districts among female workers. In the study area, a strict division of labor based on gender is observed, with male workers being responsible for land preparation and weeding, while transplanting is seen as a female labor task. Harvest and threshing are shared by male and female laborers. Market transactions in farm outputs and livestock (e.g., in terms of the shares of income from the sale of farm outputs and of livestock) appeared to be slightly more prevalent in Purulia than in Giridih. The average cash income of the surveyed farm households in Purulia was higher than in Giridih, while the reverse was true for the total imputed value of home consumption of farm outputs.

For observed household-level welfare outcomes, the estimated average per capita gross income was the same between the two districts (Rs. 4,026 in Giridih and Rs. 4,010 in Purulia), and the share of income from nonagricultural sources was roughly 40% among the surveyed households in both districts although the share of households engaged in nonagricultural activities was higher in Giridih than in Purulia.

The level of inequality in the distribution of landholding and income appears to be slightly lower in Purulia than in Giridih, as one would expect given the land reform in Purulia. The Gini coefficients for land distribution in Purulia and Giridih were 0.48 and 0.49, respectively. In terms of the estimated income distribution, the Gini coefficients for the two districts were the same at 0.38. While the observed difference in inequality between the two districts is very small, this might suggest that the land reform implemented in West Bengal was successful. Such a conclusion, however, cannot be drawn definitely without better information on the extent of inequality in land distribution across the two districts before West Bengal implemented its land reform.

Although the two districts share roughly equal levels of average per capita income and land distribution was slightly less unequal in Purulia, the estimated poverty incidence was higher in Purulia (63%) than in Giridih (57%). This is due to the higher poverty line established in Purulia than in Giridih, which itself reflects the higher average living cost in rural West Bengal.

Ownership of household appliances was slightly more common among the surveyed households in Purulia than in Giridih. For example, the proportion of the households reporting ownership of at least one (minor) appliance was 59% in Purulia and 47% in Giridih. Nearly a third (31%) of the surveyed households in Purulia reported owning a radio, while less than half that share (14%) of households in Giridih reported owning one. According to the survey, 7% of the households in Purulia and 4% in Giridih owned a television set. Eighty-six percent of the households surveyed in Purulia reported owning a bicycle, while 70% did so in Giridih.The positive correlation between landholding size and ownership of appliances appeared to be slightly more pronounced in Purulia than in Giridih.

Indicators of the level of education such as the literacy rate or the average years of schooling were slightly higher in Purulia (35% literacy rate and 3.9 years of schooling) than in Giridih (20% literacy rate and 3.4 years of schooling). Furthermore, average educational attainment appears to have risen more rapidly in Purulia in recent years than in Giridih. This can be inferred from the higher reported "maximum years of schooling within the household." Evidence of gender discrimination (e.g., differences in wage rates commanded by male and female workers) suggests that this discrimination was more pronounced in Giridih than in Purulia. There was a higher share of female-headed households among the surveyed families in Purulia than in Giridih.

Taken altogether, this comparison of welfare indicators among surveyed households in the two districts provides some evidence that, on average, households in Purulia have a higher standard of living than households in Giridih. While average levels of per capita income and agricultural productivity (measured by average rice yields) were roughly the same in the two districts, a larger number of the other welfare outcomes performed better in Purulia than in Giridih. Although the poverty incidence was slightly higher in Purulia than in Giridih, inequality indicators (e.g., land distribution and gender discrimination) suggest that Purulia is (slightly) more egalitarian. Various measures of social development suggest that Purulia has achieved a higher level of development than Giridih-individuals in the surveyed households were slightly more educated. Participation in markets, the use of modern technology and practices in agriculture, and ownership of household appliances all tended to be more widespread among the house-holds in Purulia.

Social Science and Research in Forestry

For some time, there has been a growing concern that forest managers and other resource practitioners are not using and, in some cases, are not even unaware of the research being undertaken by social scientists working in the field of forestry. In addition, there is a perception that some of the information gaps identified by practitioners are not currently the subject of research.

Social science is important to forest managers as forest management is less and less about the management of trees and increasingly about dealing with the interactions between people and forests. At the same time, there is an increasing recognition in teaching and research establishments that the old paradigm of forest science as an extension of natural sciences has been overtaken by the view of forest sciences as a major interdisciplinary subject that encompasses both the natural and social sciences.

This recognition has not yet been accompanied by major changes in the forestry curriculum or in major changes in research direction, although some groups, such as the Centre for Applied Conservation Research at the University of British Columbia, have recently expanded their mandates to include both natural and social sciences.

Many of the problems facing the forest industry in British Columbia today are related to socio-economic issues. There has been very real concern that forestry practices in the province are inconsistent with sustainable use, defined, based on the Convention on Biological Diversity as: "The use of resources in such a way that does not lead to long-term decline and maintains the potential use of those resources to meet the aspirations and needs of present and future generations." This needs to be set against a number of important trends, including demographic shifts, rapidly expanding communication capabilities, greater population heterogeneity, globalization, and, in British Columbia, an increasing recognition of the role of First Nations in resource management.

A survey of the clients of the Southern Interior Forest Extension and Research Partnership revealed that practitioners have a number of information needs that researchers are only beginning to address. These include:

- measurable socio-economic indicators;
- tools and approaches to communicate and collaborate with the public;
- management models that encourage innovation; and
- a better understanding of what people value.

This is only a partial list, based on the clients of the Southern Interior Forest Extension and Research Partnership. If the province as a whole is considered, a wider range of information needs can be identified, as shown by the research needs analysis conducted for Forest Renewal BC in 2000. This analysis indicated that provincial information gaps include:

- ways to improve citizen participation in planning;
- improved land ownership and tenure policies;
- more robust planning models;
- greater understanding of past, present, and potential roles of Indigenous peoples in resource management;

- — adaptive management of social and economic issues;
- — scale-appropriate resource management approaches;
- — indicators for measuring Traditional Ecological Knowledge;
- — ways of incorporating aboriginal values; and
- — regulation and policy frameworks that are culturally sensitive and help to foster improved relations with First Nations.

Although a number of gaps are apparent, there are also several success stories already where social scientists have helped to develop solutions to problems. These include emissions trading for atmospheric pollutants such as sulphur dioxide, conflict resolution and decision-making procedures, public involvement theory, and non-market values and benefit-cost analysis for regulatory analysis. However, for more successes in the field of resource management, more effective partnerships will be necessary between social scientists and resource managers. For such relationships to flourish, social scientists will need to change the perception of their work as "soft science" and will need to market their expertise more effectively outside their own disciplines. Specific areas where social scientists could play a role include public involvement, innovative institutions, analysis of social/economic impacts, conflict resolution, improved decision-making, certification, criteria and indicators, and social adaptive management. This is especially so given the increasing acceptance of the principles of ecosystem management in forestry in British Columbia and elsewhere.

The academic community in British Columbia and western Canada has considerable capacity to undertake social science research, particularly in the **fields of** sustainable development and communities. Many different groups are undertaking such research, although contact and collaboration between them is rather more limited than it should be. Gaps exist, including the application of economic theory to forestry, theoretical studies in general, and links between scientists and decision-makers.

Value of Social Science

Resource managers often speak of the need to incorporate public opinion, or of the need to extend results to the public. The impression given is that the general public is a single entity, rather than a heterogeneous group of people with a diversity of backgrounds, opinions, and knowledge. Social science can help sort out this diversity and structure it in such a way that a resource manager can make sense of it. Such analyses pave the way for conflict resolution between differing groups of people. Political science also has a major role to play, particularly in light of the substantial changes to forest policy that are anticipated in British Columbia in the near future. There are four key factors in the configuration of policy instruments are identified:

— formality (guidelines vs. rules)

— simplicity (number of variables)

— transparency versus clarity

— congruence

The current trend in British Columbia is towards a rules-based approach, although not everyone (especially within the forest industry) understands what this would entail. Increased professional accountability is also discussed, but again some of the implications of this appear to have been missed (such as employer-employee relationships). Developing the optimal design for institutions, processes, and policies, given the many different pressures, will be a major challenge in the coming years.

Geography is yet another field that has a contribution to make. Geographers have concentrated on a number of themes, including lifestyles and regional description, natural hazards and ecological systems, techniques such as GIS and remote sensing, policy and planning, and physical geography (geomorphology, hydrology, pedology, biogeography, and related areas of interest). A critical theme has been the interaction between humans and their environment, an area that is important for forest management and its perception. Much of this information is of value to forest managers, but remains unexploited.

Partnership Between Social Science Research and Management

There are a number of examples of successful partnerships between social science research and management. One such is the involvement of the Ktunaxa people through a process known as participatory action research in a study of the relationships between fire and black huckleberry and other plants of importance to the Ktunaxa. The study revealed that fire did indeed have an impact on the ecology of a number of species important to the Ktunaxa. Perhaps one of the most significant findings to come out of the study is the importance of listening, learning, and adapting when Western scientists undertake research in collaboration with First Nations peoples. Many of the same conclusions were reached in a study involving the Tl'azt'en Nation. In this case, the importance of trust between researchers and the community was emphasized, as was the importance of providing opportunities for the community to correct any miinterpretations that might have arisen. In both cases, Western scientists had to change their approach to meet the needs of the First Nations communities.

In another, very different example, a survey of what the public in the Arrow Timber Supply Area really feels about forest issues in the area revealed that water was the predominant concern, with ecosystem management and steady incomes also being a priority. Such surveys are complex and require special skills as the research needs to be able to relate to people so that interviews and focus groups can be conducted without leading the participants. The objectives of the research need to be clearly formulated and the research questions themselves need to be clear. This is a requirement for all research.

Another form of partnership is between organizations. Iisaak Forest Resources Ltd., a joint-venture forestry services company owned 51% by the five Nuu-chah-nulth Central Region First Nations and 49% by Weyerhaeuser, has teamed up with the Long Beach Model Forest Society. The partnership is looking at the effects of implementing the Clayoquot Sound Scientific Panel recommendations (which were strong on ecology, but weaker on social and economic issues). The findings will be used by Iisaak to improve management plans and strategies through adaptive management. Such partnerships seem to have considerable potential.

Social and Economic Information Gaps

Not all knowledge will be easily codified and analyzed using traditional Western science and statistical methods. The traditional knowledge of First Nations is an example.

This particular example also raises the issue of the relationship between Western scientists and First Nations people. Some researchers need to change their attitudes towards First Nations; there is a need for research to be undertaken with First Nations as equal partners, rather than undertaking research on First Nations. This difference may sound subtle, but it is not. Ownership of any results obtained in partnerships is something that will need to be clear from the start of any project. Forest companies need to consider social justice, environmental responsibility, and economic equity, but many do not have the capacity to do so. For those that do, communicating what is being done is a problem. This is particularly difficult in relation to descriptions of sustainability, and the health metaphor may have potential as a means of communicating information about the state of a forest. While reasonable ecological indicators exist, there are still problems in defining adequate socio-economic indicators.

The information available to community economic development practitioners is limited. Information priorities included local economic information, information on markets, and marketing and research syntheses. An apparent failure by academics in British Columbia to connect with practitioner.

Challenges and Solutions

A number of challenges were developed during the forum. Many of these relate to perception of social science. For example, why is social science seen as a "soft science" and somehow inferior to natural science? Many individuals consider that social science and social science findings are common sense. While this may be true in some instances, it is not the case for many of the problems addressed by social science. A number of solutions were identified during the meeting:

— Language was identified as a problem: There is a need for communication in plain language if interdisciplinary collaboration and researcher-management communication are to be successful.

— Better links between groups are required, especially between urban researchers and rural practitioners.

— A system of communicating research and information needs from practitioner to researchers is needed.

— New information technologies, if used, would help information access in remote areas.

— Social scientists need to be brought onto planning tables (Land and Resource Management Planning) and into mainstream decision-making.

— Social science needs to receive a higher profile at universities and in continuing education.

There is a major role for extension organizations as a clearing house for questions, as a means of transferring knowledge, as a means of facilitating exchange of knowledge, and for many other purposes.

Social science in forestry has come of age. However, it is quite difficult to identify the most appropriate next steps. There is a clear need to include more social science in forestry education. Many trainee foresters lack basic social skills such as the ability to communicate effectively or the capacity to resolve conflicts. For those already practicing forestry, there is a need to emphasize the success stories that involved social science. Most of these are typified by involving social scientists at an early stage in the process.

There is a perception amongst practitioners that social scientists have not played a major role in solving practical problems faced by resource managers. There is also a perception that social scientists do not really understand the practicalities of modern forestry in British Columbia. Again, while there may be some truth in these perceptions in some cases, they cannot be extended into broad generalizations. A sufficient number of examples now demonstrate that social scientists can provide answers to some of the important problems facing forestry. However, there is a need for better communication between practitioners and scientists so that the expertise and capacity can be better matched to the problems identified by practitioners.

Human society is undergoing changes that profoundly affect the management of natural resources. These changes include major demographic shifts, rapidly expanding communication capabilities, greater population heterogeneity, and globalization. Major demographic trends in developed countries include slowing population growth and an increase in the income, education level, and age of citizens. These trends have led to greater demands for environmental services, including preferences for conservation, environmental quality, amenities, recreation; a rise in urban interface concerns; and increasing interest in resource management by the wider population.

Various researchers have observed a phenomenon known as the "Environmental Kuznets Curve," which relates environmental degradation with per-capita income (Figure 1). The curve shows environmental degradation increasing with per-capita income until a turning point is reached, after which further increases in income lead to decreasing environmental degradation. Consistent with the Kuznets curve, the Krutilla hypothesis links increasing wealth with an increased desire for environmental quality (Figure 2).

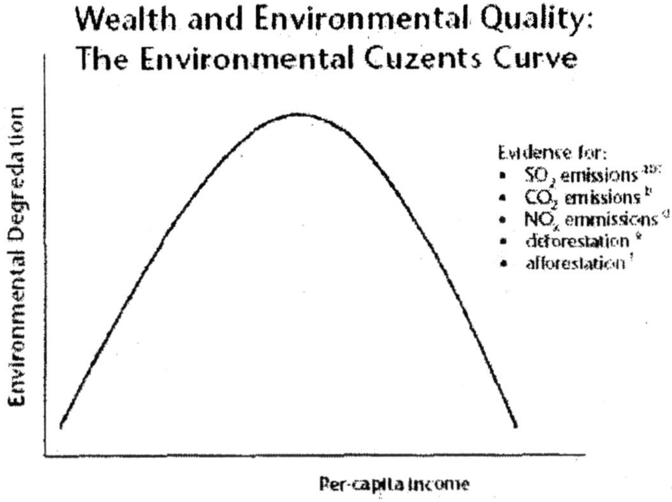

Figure 1. The Environmental Kuznets Curve.

One explanation for this link is that higher income levels make natural capital scarcer in relation to human capital, thereby increasing the "value" and demand for natural capital. Another explanation is that environmental quality is a "luxury good," and thus demand for the good will increase as incomes increase. These ideas suggest that social factors strongly influence people's preference for environmental services. These preferences, in turn, are "operationalized" through policy and management decisions, such as emission control regulations, endangered species policies, and the development of criteria and indicators of improved environmental protection.

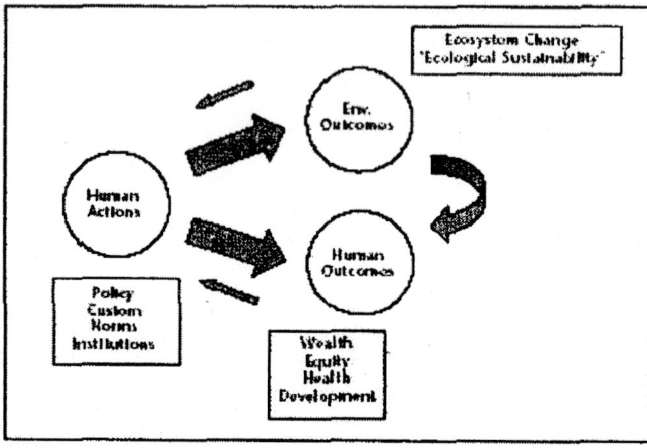

Figure 2. Krutilla hypothesis

These findings indicate the importance of social factors on resource management and on the demands placed on resource managers and policy makers. Trends in information exchange include more rapid information transmission, new delivery vehicles (Internet, etc.), new sources of information, and new scientific information (GIS, landscape simulation, etc.). However, humans have a predilection for incomplete information processing; they tend to rely on heuristics, norms, and simplifications; have trouble with ambiguities; and are susceptible to issue "framing" (i.e., simplistic labelling of events to meet personal or political ends). Thus the rapid increase in information will not necessarily lead to ease of environmental decision-making. In fact, it may lead to much more challenging situations of environmental conflict.

Social science is required to help understand how information is processed and how best to present complex information to the public and to policy makers. Trust affects the flow of information, the likelihood of conflict, and the effectiveness of resource management and policy decisions. Thus research on trust could provide valuable information to resource practitioners reliant on public acceptance of their management practices. Society is also experiencing increasing heterogeneity. The "public" is, in fact, many publics with different preferences, different backgrounds, different perceptions, and different information sources. There is an opportunity for researchers and resource managers to work together to better understand and address this diversity.

Globalization raises numerous complex social issues. Research is needed on such subjects as the linkage between local and global decisions and actions, certification, market access, supply chains, market concentration, the World Trade Organization, trade issues, global trade patterns, and changing institutions. While there is clearly no shortage of social issues that researchers can address, social scientists face peculiar challenges. Social science is perceived as a "soft science," and as such is often viewed as less credible than the natural sciences. There is a perception that specialized social science research is unnecessary because people deal every day with social issues, and can therefore make good decisions based on their personal experiences. At the same time, the complexity, uncertainty, and problems of scale in formalized social research make it difficult to translate research results into practical policy and management decisions. Nevertheless, there are success stories. Important contributions to resource management originating from the social sciences include emissions trading, non-market values and benefit-cost analysis, conflict resolution and decision-making procedures, public involvement theory, and social impact assessment methods. In conclusion, social science should play a significant role in resource management and planning. However, such involvement demands greater collaboration between social scientists and resource managers.

SOCIAL SCIENCES IN RESOURCE MANAGEMENT

The social sciences include the traditional disciplines of sociology, psychology, political science, economics, anthropology, archaeology, geography, and communications. These disciplines have fostered different sub-fields relating to resource management, each with its own traditions and terminology. Several topics are not covered by the traditional social science research disciplines (Table 1).

The categories in Table 1 are by no means mutually exclusive. On the contrary, much progress in human dimensions research (social sciences pertaining to resource management) is owed to interdisciplinary approaches. The "non-fits" category refers to important resource management concerns for which social sciences make contributions, but not any one traditional discipline claims ownership. Social science methodologies go beyond the boundaries of the traditional scientific method. Nontraditional research methods include qualitative research, participatory research (appraisal), action research, feminist research, and evaluation. Integration between the various fields and sub-fields of social science, as well as between social science and natural science, is critical. There are many challenges to such integration, however. One important concern is to ensure that the conceptualization and data collected match the same scale between social and ecological research and management concern.

Table 1.

Traditional social science disciplines and some of their sub-fields relating to resource management

Sociologists	*Social psychologists*	*Anthropologists*	*Political scientists*	*Economists*	*Non-fits*
Social and cultural impacts	attitudes	traditional values	public participation and involvement	economic impacts	wildlife–human interaction
Community cohesion	preferences and perceptions	local knowledge	conflict resolution	valuation	cultural landscape assessment
Sense of place	motivation	mapping of traditional use	alternative dispute areas	cost–benefit analysis resolution	capacity building
Community capacity	satisfaction	spirituality of wilderness	co-management	multiple accounts analysis	program evaluation
Environmental values	risk perception		evaluation of institutional arrangements	comparative analysis of economic sectors	evaluation of decision processes

The social sciences are at least as complex as in biology and ecology, yet interdisciplinary teams often include only one social scientist among numerous natural scientists. Finally, support has been inadequate for knowledge transfer and extension relating to social issues. One example of social science research relevant to resource management is a study of fairness in the forest planning decision-making process in Ontario.

One difficulty between management and social science research is the different uses of "value." Resource managers and researchers use the term very differently, especially in British Columbia. To managers, forest values are something that can be inventoried, while to social scientists, values are something that people have. The challenge is to incorporate social research contributions meaningfully in ongoing and newly emerging management issues such as certification, the development of criteria and indicators, and public participation processes. Human dimension research needs to be taken beyond mere rhetoric. There must be more meaningful integration of social science with forest management, operations, and forest policy.

Social science at the University of British Columbia (UBC) includes the Forest Economics and Policy Analysis (FEPA) program, which examines forest economics and policy perspectives relating to Canadian forestry, and the Sustainable Development Research Institute (SDRI). Researchers focus on forest sociology, certification, tenure, public participation, climate, and community stability. Forintek, Canada's wood product research institute, conducts research on forest product markets and trade. The University of Victoria addresses environmental law and policy and environmental perspectives on sustainability. The Pacific Forestry Centre (Canadian Forestry Service) has supported research on the economics of silviculture, certification, secondary manufacturing, climate change, and wood supply. Royal Roads University has formed links to SDRI at UBC and also focuses on non-timber forest products. The University of Northern British Columbia's program includes natural resource sociology, sustainable communities, cultural and social issues including First Nations, and tourism. Simon Fraser University (SFU) contains the Cooperative Resource Management Institute and the Community and Economic Development Centre; the latter focuses on forest-based communities. Also, SFU houses the Centre for Tourism Policy and the Forest Economics Program.

The University of Alberta hosts the Sustainable Forest Management Network and supports research on non-market valuation, tenure, the economics of silviculture, forest sociology, First Nations, sustainability, climate change, political science, and economic impacts. The Northern Forestry Centre (Canadian Forest Service) conducts research on forest-dependent communities, regional economic impact modelling, non-timber valuation, forest sociology. attitudes and values; and economics of silviculture, tourism, and ecotourism. Researchers at the University of Calgary study non-timber valuation and climate change.

Social Groups and Forest Values

Natural scientists tend to study their own problems while social scientists merrily work in isolation on separate problems. There is usually little to no interaction across domains. In true interdisciplinary projects, social scientists need to be involved in formulating the problem and not only involved after the fact. Often, social scientists are involved for education and public relations purposes for problems that are defined by natural scientists.

Many people think social science and social science research findings are common sense. For any given problem, usually at least half a dozen contradictory "common sense" hypotheses can be offered. After the fact, whichever hypothesis is demonstrated to be "correct" will seem like common sense to many people. However, we will not know which hypothesis is correct unless we obtain empirical data. Many graduate students who are interested in researching "social forestry" problems do not have basic training in the social sciences; involving them in forestry social science research is time-intensive. This also applies to involving community members in research. There is not enough emphasis on basic (theoretically oriented) social science research on forestry issues. There is a bias in funding for forestry-related social science topics related to timber production. This does not facilitate research that is more theory-driven.

Policy objectives for research on public values and attitudes are unclear. What role should research on public values and attitudes play in forest management? It is important to design research that enables researchers to capture intergroup differences. In forestry circles, there is a tendency to talk about the general public as a single entity. More emphasis needs to be placed on values, attitudes, and opinions about forestry and exploring differences among groups. We can't talk about finding out "what the public wants" when we talk about sustainable forest management. "The public" is not homogeneous-they value different things about forests. We have to consider the opinions of different social groups. Values in social science are cultural ideas of what is desirable, right, and appropriate. Values can only be determined from people's thoughts and behaviours.

In a recent study called Measurement of Forest Values and Attitudes toward Forest Management funded by Forest Renewal BC and led by David Tindall, differences in priorities among nine types of values among 13 different groups of people were analyzed. These groups are not mutually exclusive (e.g., someone can be both a scientist and an outdoor recreation enthusiast). Questions were grouped according to the following nine values:

— Ecological and environmental
— Science and education
— Equity
— Community sustainability -
— Aesthetic
— Economic
— Work
— Outdoor recreation

Cultural Survey respondents were divided into the following 13 groups:

— Artists/writers
— Community representatives
— Educators
— Environment and parks (government employees)
— First Nations
— Forest industry
— MOF managers, employees

— Environmental organizations
— Recreation groups
— Scientists
— Tourism/recreation operators
— Trappers/ranchers
— Unions

Twelve out of 13 groups ranked ecological and environmental values as most important. Another interesting conclusion is that value rankings relate to the relationship of the group to the forest. For example, community representatives rate community sustainability as most important, whereas the forest industry ranks economic values as second most important after ecological values. Notably, the Ministry of Forests respondents also ranked economic values as second most important. Respondents from environmental organizations ranked economic values as least important, as did most other social groups.

Artists and writers rated aesthetic values as second most important after ecological values. Differences in value ranking between people involved in the field of forestry, and non-foresters were compared. Foresters ranked economic values as second most important, whereas non-foresters ranked economic values as least important. There tended to be two conflicting worldviews: that of people concerned primarily with economic values, and that of people concerned with ecological, recreation, and aesthetic values. Therefore, much of the conflict between groups regarding forest land use and management is not based on disagreements about specific practices, but rather it reflects their conflicting worldviews. The results of this research support the importance of forestry professionals determining the values and preferences of others, as it is likely that their values will be different from those of people employed in the field of forestry.

Institutional and Policy Design

How can political science knowledge be made more relevant and practical to forestry? The policy regime framework can be used to try to understand why policy makers make the decisions they do (Figure 3). The process goes from agenda setting to decision-making. The product of this is policy contents, such as the Forest Practices Code. This policy resulted from an interaction of strategic actors: environmental groups, industry, politicians, and registered professional foresters. Each group has different interests and resources. The interaction between these groups influences the policy outcomes. It is important to recognize the larger context of the electoral system and how this affects policy. The new provincial government in British Columbia elected in 2001 will be the most important factor affecting forest policy outcomes in the next few years, but local and global markets, and public opinion also play an important role.

In the 1990s there was a moderate change in the areas of timber supply, land use, forest practices, jobs, and pricing. This is surprising since there had been a high level of activism and public pressure on forest policy in the last decade. Table 1 shows factors that promoted change and that worked against change. The newly elected government has proposed changes to the Forest Practices Code. The experience of the previous New Democratic government can help us understand how the new government may or may not be able to succeed with their proposed policy changes.

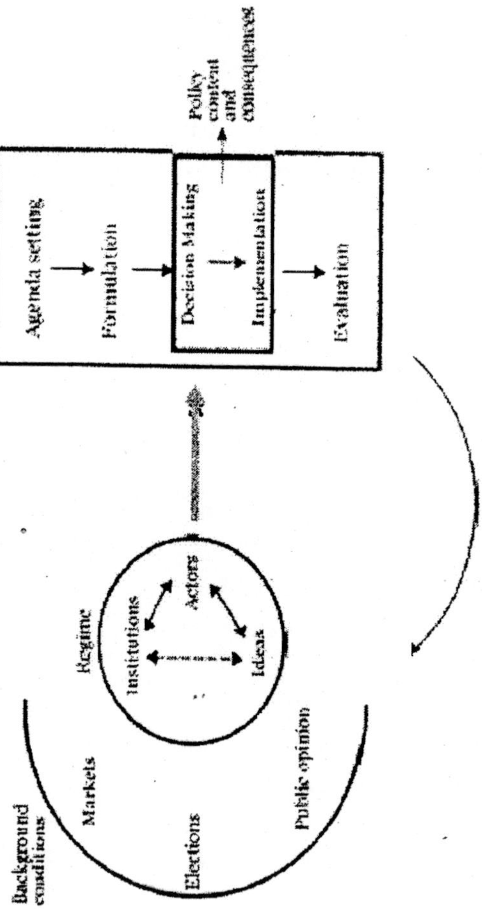
Agenda setting
Formulation
Decision Making
Implementation
Evaluation
Policy content and consequences
Background conditions
Markets
Elections
Public opinion
Regime
Institutions
Actors
Ideas

Figure 3

Institutional design also affects forest policy. Organizations have different biases. For example, the Ministry of Environment, Land and Parks tends to be more environmentally friendly than the Ministry of Forests. Therefore, who is authorized to make a decision can affect what the outcome will be. Which ministry is put in charge of reforming the Forest Practices Code will greatly affect the Code's direction. Structural forces affecting forest policy include environmentalism, globalization, intensification, advances in science and technology, certification, and reconciliation with First Nations. People who live farther from the site of logging activity tend to favour conserving the forests. We now have a drive towards decentralization and community forestry. It is questionable whether this will make forestry more environmentally friendly.

Collaborative decision-making has become fashionable in today's policy development. However, we need to determine the appropriateness of public involvement in different policy processes. Four key factors are involved in the configuration of policy instruments:

— formality
— simplicity
— transparency versus clarity
— congruence

How do we make rules for forest management that can cope with the diversity of ecosystems and communities within which they are applied? One option is to allow professional discretion in applying the rules. This is not transparent though, as different professionals may interpret the same situation differently. Another option is to create a "cookbook" for each situation. This is transparent, but too complex, as there are so many different possible situations.

We are in search of the optimal balance between these two options. Discretion in the application of rules allows for flexibility and innovation, yet it creates a dilemma for transparency and accountability. The core challenge lies in developing optimal design for institutions processes and policy.

Geography and Natural Resource Decision-making

Geography has played an important role in resource management throughout the last century. What is geography? People tend to think of it as just mapping. However, geography encompasses many different fields: physical geography, human ecology, aerial differentiation, and spatial science. It is both a physical science and a social science. Human ecology started in the early 1900s. It is the study of the relationship between humans and their environment. It emphasizes the natural environment and promotes the idea of environmental determinism (the natural environment controls one's behaviour). Human ecology developed a bad reputation, as it promoted ideas related to environmental determinism-for example, "people in the tropics are lazy, because it's too warm to work there."

Aerial differentiation became a major focus in geography in the 1930s. It is a region-by-region analysis, such as a study of the major natural regions of British Columbia. In the 1950s, geography went through a quantitative revolution. Quantitative research methods were becoming available, and geographers started to analyze quantitatively the distribution of phenomena in space. In the 1960s, as people began to perceive an environmental crisis, the study of human ecological systems became important again. Human ecologists looked at how people respond to extreme geophysical events such as floods or tornadoes. They asked questions such as, "Why do people live on floodplains?" Geographical information systems (GIS) and remote sensing developed in the 1960s. Mapping resources became important. The Canada Land Inventory was conducted to map all of Canada by landuse type (agriculture/forestry/soil types, etc.). The world's first GIS was developed in Canada in 1963. Once GIS was available, researchers and practitioners needed to gather more data for analysis, so the field checking of remote sensing developed. First aerial photographs were used, then infrared photographs, and later satellite images.

Environmental impact assessment practices developed in the late 1960s due to continued interest in environmental issues. The U.S. Environmental Policy Act of 1969 spurred the growth of policy studies in geography. The importance of physical geography such as geomorphology also increased during this period. The interaction of humans and the natural environment was being studied intensely. Geographers have been involved in the following areas:

— lifestyles and regional description

— policy and planning

— natural hazards and ecological systems

— physical geography

— techniques (GIS, remote sensing)

In the 21st century, "geographer" is a rare title for natural resource professionals; however, people trained in geography can have an important role in resource management. Geographers can help in data collection, analysis, and synthesis, and they can clearly communicate written results.

Residents, stakeholders, and sustainable forest management

The province's Innovative Forestry Practices Agreement (IFPA) Pilot Program has six pilot projects throughout British Columbia and involves both resource management and social aspects. The pilots are partnerships between industry, government, specialists, and stakeholders and are established under provincial legislation to create sustainable forest management programs. Arrow IFPA is one of the pilots and is focused in southeast British Columbia.

The framework of the Arrow IFPA is still in the developmental stage. Work on establishing sustainability criteria and indicators, setting appropriate scales and thresholds, investigating desired future conditions, and determining impacts of natural disturbance and practices is underway. Monitoring and modelling are key components of the framework. Another key component is setting achievable and concurrent management objectives for social, ecological, and economic values. The Arrow IFPA is gathering technical and scientific information on each of the values. This information will then be taken to residents and stakeholders.

Arrow IFPA has undertaken a social survey and will be completing a stakeholder analysis. The survey was used to determine how the public feels about forest resource issues in this area. Surveys were randomly mailed out to people in the Arrow Forest District and residents within the community of Nelson. Of the nine resource values tested, "water" came out as the top concern, although respondents considered all issues important. The surveys asked participants to map priority areas and demographic information.

The survey is a first step in determining how residents and stakeholders view management of the surrounding forests. The next step will be to complete a stakeholder and multi-criteria analysis. As part of the IFPA's monitoring plan, a follow-up survey will be undertaken over the next few years. Although it is unlikely that consensus will be achieved in the Arrow TSA for all management options and values, it is important to aim for it. Generally, more emphasis on social science will be used in the development of concurrent management strategies.

Public attitudes towards natural resource management

Good social science research is about finding out what people think and how many think it. To adequately measure public attitudes towards natural resource management, some basics are important:

— Careful selection of the researcher. It needs to be someone who is capable of relating well enough to people that they can carry out good interviews or focus groups.

To measure attitudes towards land-use planning processes (Commission on Resources and Environment [CORE] and Land and Resource Management Planning [LRMP]) in resource-dependent communities, a mixed strategy was devised combining quantitative and qualitative methodologies. Interviews with key people in each community were conducted. Local officials, people involved in the CORE and LRMP processes, people interested in resource management, women, young people, and First Nations were interviewed. Research assistants remained a month in each community so that they could get a feeling for each community. The interviews were followed up with a series of focus groups in each community. Finally, a mail survey was used to determine the breadth of the opinions. The researchers felt that it was important to return the research to the community. They held an open house in each community and presented their research findings. This also allowed them to confirm their findings.

Considerable time was spent developing a relationship of trust with the community and using a community liaison to help with arrangements. This ensured that the community had someone they trusted involved in the project. The first project was started with extensive interviews within the community, including elders, youth, those involved in the tree farm licence, band officials, and others. To help document the information that was uncovered, a videotape was created. In the second project, interviews were also used, but followed up with focus groups to ensure elders and youth felt comfortable sharing their opinions.

6

Prospects of Agricultural Baseline

Macroeconomic assumptions represent expected future trends in key variables, but exclude any variations due to business cycles. Supply projections assume average weather conditions in each year. Economic and agricultural policies are assumed to continue to evolve along recent trends, based on analyst judgment.

World trade in most major bulk agricultural commodities is projected to expand more rapidly during 1995-2005 than during the 1980's or early 1990's. Trade in grains, particularly coarse grains, is projected to grow the fastest among bulk commodities. These gains are driven largely by projections of stronger economic growth in developing regions, including China, Asia, Latin America, North Africa, and the Middle East. In these regions, rising incomes are leading to diet diversification, rising meat demand, expanding livestock sectors, and higher demand for feed. Wheat trade is also projected to increase, due to strong global demand growth. Combined trade in soybeans and meal will strengthen, a result of the same expansion of developing country feed-livestock sectors that will push up coarse grain trade. Growth in soybean oil trade is also projected to be faster than in the 1980's, but will remain slower than some competing oils because of its high relative price. Raw cotton demand and trade is projected to be stronger than in the early 1990's, but not match the 1980's when there was increased substitution of cotton for synthetic fibers.

The generally favorable world economic outlook is expected to spur growth in meat demand and trade. Additional impetus is expected from already negotiated reductions in trade barriers, primarily in East Asia. Meat demand growth is projected in several countries in the Pacific Rim and Latin America, with the Pacific Rim providing the most growth in both consumption and import demand. Rapidly increasing incomes in China, Taiwan, South Korea, and a number of other countries in the region stimulate demand for meat. The United States is well positioned to provide a variety of meat products to these markets. Growth in meat import demand in the Former Soviet Union (FSU) is projected to slow. Although declines in meat consumption will slow and demand will turn upward after 2000, domestic FSU production of meat is also projected to begin increasing. This could reduce the region's dependence on imported meat, although the United States is expected to continue to supply low-priced parts and trimmings to that market. The value of U.S. meat exports is projected to grow an average of about 4 percent annually during 1997 to 2005, somewhat slower than the rapid ascent of the past several years.

Agricultural Price Projections

Along with relatively strong growth in trade, the baseline projections indicate tightening markets for the major bulk commodities. Projected prices for the major commodities will continue to decline in real terms through 2005, but at a slower rate than long-term trends. Strengthening U.S. and global crop prices stem from the projected growth in demand, combined with the outlook for somewhat slower growth in yields. Although crop area is expected to rise along with price incentives, yield gains tended to slow for major crops in a number of regions during 1985-95. Although yields are expected to show some response to price incentives, the extent to which global supplies will respond in an environment of firmer prices is a key uncertainty in the outlook.While firmer real crop prices are projected, meat prices are expected to continue to decline roughly consistent with their long-term trend. Particularly in the United States, the impacts of higher feed prices are expected to be offset by continued efficiency gains associated with improved feeding practices and vertical coordination in the meat industry.

Major Uncertainties

The trade projections are sensitive to the assumptions for continued strong economic growth in developing Asian countries, combined with strengthening growth in Latin America, North Africa, and the Middle East. The price and trade projections are closely linked to projected gains in crop productivity. The responsiveness of yields to an environment of firmer prices and increasingly market-oriented policies, as well as improved investment conditions in many developing countries, is uncertain. More specifically, the projections are sensitive to the uncertain outlook for supply, demand, and policy developments in China, the EU, and the transition economies of the FSU and Central and Eastern Europe (CEE). Prospects for the huge China market are uncertain because rapid growth and reform have greatly complicated assessment of future policies and technical supply and demand coefficients. The EU trade projections depend on assumptions regarding the nature of policy adjustments that may be undertaken to comply with the Uruguay Round Agreement, and on future supply response. And, although the FSU is projected to have a reduced role in world grain trade, it is inherently difficult to accurately assess future policies and economic relationships in the transition economies of the FSU and CEE.

International Macroeconomic Assumptions

Estimates for macroeconomic variables through 1997 were the most likely short-term forecasts of economic growth, inflation, and financial market behavior at the time the macroeconomic assumptions were prepared in October 1996. The forecast for 1998 is a transition between the short-term forecast and the long-term projections.

The long-term projections for the macroeconomy for 1999-2005 reflect trend assumptions for some indicators combined with standard relationships between major macroeconomic variables. The absence of business cycles beyond the first or second year of the forecast reflects a conviction that business cycles, as well as shocks to the macroeconomy like large oil price increases, cannot be accurately forecast. This macroeconomic setting avoids distorting the long-term baseline for agriculture that would result from introducing unpredictable swings in macroeconomic variables.

The world macroeconomic outlook continues to be favorable over the next decade, with global real GDP growth averaging about 3.0 percent, well above growth during the first half of the 1990's. In the developed countries, macroeconomic growth is projected to average about 2.5 percent through 2005. Europe.Over the next 2 years, Germany, France, Italy, and the United Kingdom, along with the other countries of Western Europe, will consolidate their recoveries from 1996's growth recession and will show faster growth. From then through about 2000, they will be in a more mature phase of their business cycle recovery, with slowing growth, but no recession. This growth will take place in a low-inflation environment (German inflation is currently 1.5 percent), since growth will be modest, and expected labor market conditions (unemployment around 10 percent) will retard wage growth. Through 2000, fiscal spending will be restrained, driven by the Maastricht Treaty's deficit-to-GDP requirements. Monetary policy over the same time period keeps short-term interest rates only modestly higher (about 50 basis points) than their current levels. During 2000-2005, European economic growth is expected to settle into the mid-2 percent range, while inflation stays around 3.0 to 3.5 percent.

Unemployment will remain at or near double digits in many countries, holding down wage growth and consumption spending. Additionally, higher taxes to reduce fiscal deficits to acceptable levels for monetary union will further dampen consumption growth. Investment growth looks to be the strength of many European economies, but will not reach levels of the late 1980's. In part, this is due to uncertainties regarding monetary union as it conflicts with domestic needs.

The baseline projections assume that at least limited monetary union will occur in Europe on schedule, but the actual pace of unification is the principal source of uncertainty. There is a gathering momentum for monetary union to occur on schedule with more countries participating in the first wave. Spain, Portugal, and Italy are pressing to be included in this group, even though they will not strictly meet all requirements. Early entry of these countries changes prospects for the U.S. dollar's value. Currently the baseline has the U.S. dollar showing near-term strength, with slight weakness after 2000. If participation in monetary union is widespread, then European monetary policy would have to be more accommodative, and the dollar would appreciate throughout the forecast period.

With monetary union increasingly likely to occur on schedule, governments will have to install fiscal and tax policies that move government deficits toward the Maastricht limits earlier than some have expected. For the marginal countries, this could lower growth expectations significantly. Monetary decisions by central banks will continue focusing on stabilizing currency values against the German Mark, but marginal countries intent on being among those initially included in the union will have to pursue stricter policies.

Again, this could result in slower growth than assumed in the forecast. Japan:Japan's near-term prospects have improved somewhat. Overall, economic signals are mixed for the near term, although the weakened Yen, low interest rates, and improved consumer sentiment will help lead Japan's economy out of recession and help maintain modest 2.0-2.5 percent annual growth in the medium term. Japan's banking crisis is expected to hold lending and investment spending below historic levels.

Unemployment will be at historic high levels, with companies yet to expand hiring. Labor markets will only slowly improve, and therefore personal savings will remain high. Recent fiscal policy has been stimulatory, but the government will soon show restraint as it seeks to lower the recession-era deficit. Low inflation expectations, though, will allow long-term monetary policy to be accommodating, with real interest rates in the 1 to 2 percent range. Overall, fiscal and monetary policy will be slightly stimulative, but Japanese trend growth of about 2 percent will still be lower than previously seen.

The lower growth is partly due to Japan's rapidly aging population and the resulting impact on labor force growth. Investment growth will also be slower than previously, reflecting low capacity utilization-a remnant of the investment boom of the late 1980's and early 1990's-higher costs of investment, and lower corporate profits. Also, high unit labor costs, caused in part by lifetime employment policies, will push more investment offshore. With an inflation outlook that is low relative to other major economies, the Yen should strengthen slightly, with import growth outstripping export growth, toward the end of the forecast period.

Canada: Canadian fiscal policy will be tight for the next several years as the Federal budget moves to balance and Provincial budgets run modest surpluses. In 1995, Canada had the highest debt-to-GDP ratio in the G-7. Yet, in 1995 and 1996, Canada made better progress in reducing its structural budget deficit than any other G-7 country. Short-term interest rates have been 1 percent below comparable U.S. rates, and Canada registered a current account surplus in the second quarter of 1996-both events not seen since the 1970s. Uncertainty about the Canadian dollar has subsided as Quebec postponed the next secession vote to past 2000. Because of a good inflationary environment, and the favorable political and fiscal policy climate, the Bank of Canada (BOC) will continue to keep short-term rates low.

GDP growth in 1996 will be less than 2.0 percent, largely due to a weak first half. But, Canada should see abovetrend growth for the next 4 years. This is despite expected moderate U.S. growth over the same period. Continuing recovery in real wages and low short-term interest rates will bring low double-digit construction spending growth. These factors also result in more consumer durable spending, particularly since Canada has an aging fleet of cars. The expected strong equipment spending stimulated by low interest rates will largely offset the negative impacts of lower government spending on domestic growth. As the Canadian economy has substantial slack, inflation should be below 2.0 percent for the next 2 years, and then accelerate slightly.

Low inflation and continued political stability will allow a gradual appreciation of the Canadian dollar. That movement, however, will be slowed by the low interest rate policy of the BOC. With the second lowest unit labor costs among the G-7, and expected moderate growth in its trading partners, net exports will be strong. With GDP growth the strongest of all developed economies, unemployment will move well below its current rate of 10.0 percent.

Over the longer term, NAFTA and declining long-term interest rates should stimulate investment spending and exports. Additionally, the competitiveness of the Canadian dollar (remaining below the U.S. dollar in value), will further support export expansion. Unemployment and wage growth will also improve as productive capacity and capacity utilization rise. With the steady improvement in Canada's economic environment, growth is expected to be nearly 3.0 percent in 2001 to 2005, giving it the best long-term outlook among industrialized countries.

Transition Economies

After 5 years of economic decline, gains in real output have lifted per capita GDP over levels before market reforms began in most of the CEE region, particularly in the northern tier countries. Lagging reforms in the FSU have stalled its long-awaited recovery until 1999. Regaining former per capita GDP levels in the FSU is further in the future. Reducing inflation was critical in halting and reversing the output contractions that initially characterize transitions from central planning. While mid-term prospects reach up to 5 percent average GDP growth, the long-term economic outlook is somewhat slower.

Stagnant population growth and a rapidly aging labor force mean lower long-term growth than in other emerging market countries. Former Soviet Union.Market reforms are continuing to transform these formerly centrally planned economies, but at varied speeds and extent. The largest economy, Russia, will be among the first to post positive economic growth, but not until late in the decade. The transition to a market system has been slow and protracted. The financial sector, such as the stock and bond markets in these countries, had to be created practically from scratch. The allocation of domestic saving into private investment is limited by slow development of a modern capital market. The banking system had to be restructured, particularly in those countries that introduced new national currencies. After almost a decade of reforms, market-based pricing, and continuing privatization, the economies of the FSU will begin to grow again by the end of the decade. Over the next decade, aggregate growth of about 3 percent is projected for the FSU.

A more optimistic outlook will depend on the extent of privatization and foreign ownership, how much current fiscal imbalances are reduced, and how soon exchange rates stabilize. Inflation will remain in double digits but on a downward trend. The expected strong demand for capital imports will bring the current account into deficit. The region's comparative advantage in natural resource-based exports will provide much of the earnings needed to finance the capital imports. More foreign investment should be forthcoming to develop these resources.

Central and East Europe.This group of transition economies initiated market reforms earlier and to a greater extent than the former Soviet Union and are now reaping the economic rewards of fast growth. Average GDP expansion of 5 percent is expected over the next few years, and 4.5 percent from 2001 to 2005. Like other developing countries, gross fixed investment will be the fastest growing component of total domestic demand.

Inflation will remain relatively high in the mid-term before falling to single digits after the year 2000. More stable exchange rates and improved terms of trade are projected over the coming decade. Direct foreign investment into the region will continue to be strong as these countries become more economically and politically integrated with the European Union.

As in Asia, the key to sustained high growth of 5 percent or more is trade-within the region and with the EU and the FSU. Direct foreign investments follow naturally from open trade, as it has in the northern tier countries of Poland, Hungary, and the Czech Republic. This cycle of trade and investment will lift per capita GDP in the area, which currently approximates that of South America. Compared with 1990, the year after market reforms generally started, Central and East Europe has recovered output lost during the recession years in the early 1990's.

By year 2000, per capita GDP will be 17 percent higher than when the transition began.

In the southern tier countries-Bulgaria and Romania-market reforms were initiated more slowly or much later. As a result, foreign direct investment has been minimal. As in the FSU, privatization has lagged that in Poland, Hungary, the Czech Republic, and the Baltic republics. Exchange rates have continued to depreciate, undermined by low foreign exchange reserves. Financial conditions will only slowly improve. As a result, output growth is lagging behind rates in Poland, Hungary, and the Czech Republic. However, because output was initially low, output growth in the next few years will be able to catch up with current growth rates of those northern tier neighbors.

As markets and competitors for U.S. agricultural exports, the growth of developing countries significantly influences global agricultural trade. Led by Asia, aggregate growth over the next 10 years is projected to average about 5.5 percent, somewhat faster than over the past decade. While Asian growth may slow somewhat, growth prospects in other developing regions are improving. Freer markets characterized by less price control and fewer trade barriers, more stringent fiscal and monetary policies, and the phase-out of artificially controlled exchange rates are assumed for many of these economies.

Mexico: A growth recovery in 1996 after 1995's deep recession is expected to put Mexico on a path of 4.2-percent average annual growth over the mid-term. Longerrun growth prospects are brighter at 5.5 percent as the economy returns to full employment and to the consumption and investment patterns that held before the December 1994 peso devaluation. This means a gradual appreciation of the real exchange rate, moderate inflation rates of less than 10 percent, and domestic investment growing at more than twice the pace of private consumption. NAFTA will generate or enhance the chances for expanded trade volume, the return of previous levels of foreign direct investment, and restored purchasing power in Mexico.

Beyond 2000, the projections assume sustaining previous gains, with exports of goods and services growing slightly faster than imports. The terms of trade for the projections are favorable, although the current account balance remains in deficit, reflecting the continued importation of foreign financial capital. The gradual appreciation of the real exchange rate (in pesos per dollar) means that the nominal pace of depreciation will not keep up with the domestic rate of inflation. This is largely the reason for the slowly increasing trade deficit through the forecast period. Thus, Mexico will likely run a trade deficit with the United States, its largest trading partner.

A great deal will be required for this scenario to be realized. The economy's basic problem is a low savings rate. The previous dependence on foreign capital left the peso and interest rates vulnerable and volatile. Growth and increasing productive capacity will be difficult in this situation because the government will have to run a budget surplus to help raise national savings, and keep monetary policy tight to keep inflation down and to discourage an outflow of funds. Prudent macroeconomic policies will be required for Mexico to avoid currency crises that push up the cost of foreign debt service, raise capital import costs, and slow economic growth.

South America: Strong growth is projected for South America, led by the MERCOSUR core countries of Brazil and Argentina. Freer trade will further integrate the economies of these countries as they move toward eventual hemispheric free trade with NAFTA countries. Behind the strong output expansion is intra-regional trade as well as heavier foreign direct investment in anticipation of trade integration with the United States, the world's largest market. Recent market-oriented reforms and larger private sectors are also behind the region's better prospects. The past environment of overvalued currencies, large trade deficits, fiscal deficits, and low internal savings and investment is expected to improve. New macroeconomic policies now permit lower inflation and more competitive industries as import barriers fall.

China: The Chinese economy will maintain the strongest growth in Asia over the next 10 years, averaging almost 9 percent in the next 5 years and 8 percent in 2001-2005. With population growth slowing to an average 0.8 percent annually, per capita GDP growth will be at least 7 percent per year. Consumer price inflation will remain in double digits throughout the period, and because the nominal exchange rate will depreciate at a slower pace, the real exchange rate will appreciate. Inefficient state-owned enterprises will continue to burden the government's fiscal well-being. However, the country's high savings rate limits the impact on private investment.

Trade volume and investment flows are likely to grow even more as the Yuan becomes fully convertible before the end of the century. Trade competition with its fast-developing neighbors also implies a broadening of industrial technology and less dependence on labor-intensive industries. East and Southeast Asia.Although growth in South Korea and Taiwan is expected to be somewhat slower than during the last 10 years, these and neighboring Southeast Asian economies are expected to maintain relatively high rates of income growth during

1997-2005. The economies of this region are the most integrated among the developing countries-by trade and intra-regional investment. In particular, driven partly by direct investments from Japan, industrial development is fast approaching that of the developed economies as measured in per capita GDP. Overall output growth will settle down to a more sustainable pace as domestic investment decelerates. The level of trade competition and invested technology in the area is such that graduation to higher-value products appears to be essential in maintaining trade competitiveness.

Trade relationships promoted through the Association of Southeast Asian Nations (ASEAN) and Asia-Pacific Economic Cooperation (APEC), as well as appreciating currencies, are increasingly important factors in longer term prospects for trade and growth.

South Asia: While projected growth is not as impressive as in East and Southeast Asia, output expansion will nonetheless result in overall per capita gains of about 3.6 percent in South Asia through the next decade. Per capita GDP in year 2000 will be a third higher than in 1990. Greater trade and investment inflows will fuel the region's growth and, because of the region's sizable population, internal demand will sustain healthy economic activity, much like China's growth pattern. While India's growth is strong, double-digit growth like China's is unlikely. Nevertheless, given the size of potential market demand among India's neighbors, as well as its proximity to the oil producers of the Middle East, the outlook for India's export growth is bright.

Africa and the Middle East: Countries in this region are projected to start achieving positive per capita GDP growth, after contracting in the first half of the 1990's. Increases in the real price of crude oil help the Middle East and North Africa. Per capita real GDP gains in Sub-Saharan Africa, however, will be close to zero as economic growth strains to keep up with population growth. The two largest economies in Sub-Saharan Africa, Nigeria and South Africa, will have contrasting per capita growth paths-negative and positive, respectively.

Agricultural Policy Assumptions

United States

The baseline reflects provisions of the Federal Agriculture Improvement and Reform Act of 1996 (1996 Act), which was signed into law on April 4, 1996. The 1996 Act is a milestone in the evolution of U.S. agricultural policy because it fundamentally redesigns income support programs and discontinues supply management programs for producers of wheat, corn, grain sorghum, barley, oats, rice, and upland cotton. The 1996 Act replaces a system of deficiency payments, based on the difference between a pre-set target price and the higher of the market price or the loan rate, with a system of fixed production flexibility contract payments that are largely decoupled, since there is virtually no link between payments and current plantings. The 1996 Act encompasses a wide range of issues related to agriculture, including commodities, trade, conservation, nutrition assistance, agricultural promotion, credit, rural development, and research, extension, and education. Major changes related to production agriculture are in the commodity provisions, the agricultural trade provisions, and the conservation provisions of the 1996 Act.

The most important impacts result from policy changes in four main areas covering income-supported crops, price-supported commodities, agricultural trade, and conservation, and are summarized in the box ("Four Areas of Policy Change in the 1996 Farm Act"). For more detail on the U.S. policy assumptions, see Agricultural Baseline Projections to 2005, Reflecting the 1996 Farm Act. A crucial factor affecting land availability in the United States is the assumption regarding the Conservation Reserve Program (CRP). Over 20 million acres of CRP contracts expire in 1997.

CRP enrollments in 1997 are assumed to keep the CRP from falling below 30 million acres. Enrollments in subsequent years are assumed to gradually increase the CRP to over 36 million acres by 2001. Most land enrolled in the CRP is in areas traditionally planted to major field crops, thus limiting the response of planted acreage to rising prices and net returns.

Policy assumptions underlying both U.S. and foreign projections are based on full compliance with all bilateral and multilateral agreements affecting agriculture and agricultural trade as of January 1997. Bilateral agreements affecting agricultural trade between the United States and Canada, the United States and Mexico, the United States and Japan, and the United States and Korea are examples of recent agreements for which full compliance is assumed. In contrast, no compliance is assumed for any agreements under discussion or not formally ratified by January 1997. In the multilateral sphere, the projections assume full compliance with the internal support, market access, and export subsidy provisions of the Uruguay Round Agreement on Agriculture by all parties to the agreement. Several potential multilateral agreements that could have a significant impact on agricultural trade are now under consideration, but are assumed not to occur in these projections. These include:

— No accession to the World Trade Organization (WTO) by the FSU, China, or Taiwan;

— No enlargement of the EU-15 to add one or more Central or East European countries;

— No implementation of more liberalized trade among the Asia-Pacific Economic Cooperation (APEC) countries, and;

— No expansion of NAFTA to include additional countries.

Domestic agricultural and trade policies in individual foreign countries are assumed to continue to evolve along their current path, based on the consensus judgment of regional and commodity analysts. In particular, the process of liberalizing economic and trade reform underway in many developing countries is assumed to continue. Similarly, the development and use of agricultural technology and changes in consumer preferences are assumed to continue to evolve based on past performance and analyst judgment regarding future developments. Key assumptions underlying the projections for major foreign countries are summarized below.

European Union

The baseline projections for the European Union (EU) incorporate policy changes adopted as part of the 1992-93 reform of the EU's Common Agricultural Policy (CAP), as well as EU commitments under the Uruguay Round agreement that limit subsidized exports and improve market access. The final step of planned price cuts under

CAP reform took place during 1995/96. Basic support prices are assumed to remain at 1995/96 nominal levels for most commodities, but internal market prices may be driven below support levels in order to clear domestic markets. If Uruguay Round limits on subsidized exports are binding, excess supplies will have to be absorbed on the internal market, driving market prices down.

The annual set-aside program instituted for grains, oilseeds, and protein crops is assumed to remain in effect, with the set-aside rate being used as a policy instrument to adjust production to market conditions. The baseline assumes that the EU's Uruguay Round commitment on internal support is not a binding constraint, since many policies resulting from CAP reform meet the World Trade Organization (WTO) "production-limiting" criteria and are exempt from reduction commitments.

Tariffication of nontariff barriers and tariff reductions are assumed to have little impact because the high tariff equivalents established for most products are unlikely to permit significant additional imports. Continued high levels of import protection mean that price transmission from the world market will be negligible for all baseline commodities except oilseeds and products and, in the later years, wheat, rye, and oats. The most important Uruguay Round commitments for the baseline are the limits on subsidized exports and the minimum import levels agreed under the market access provisions.

Major uncertainties include what measures the EU will use to meet its subsidized export and minimum import commitments within the limits of the Uruguay Round, and what measures the Commission will adopt to deal with the projected imbalance between beef production and consumption in the wake of the bovine spongiform encephalopathy (BSE) crisis. The baseline assumes that the EU will use current policy mechanisms to meet its Uruguay Round limits on subsidized exports. For grains, it is assumed that any production in excess of domestic use that cannot be exported will depress the internal market price and dampen output. The EU will use the set-aside rate to constrain surplus production.

The setaside rate is 5 percent for 1997/98, and is assumed to move up to 15 percent in 1998/99 and then remain at 12 percent for the rest of the baseline. In the longer term, the baseline assumes that the EU will not increase intervention purchases and accumulate stocks beyond the historical average level-accumulation of intervention stocks is viewed as a short-term strategy for dealing with excess grain supplies. The baseline assumes that the EU will export grain without subsidy only when the world price is equal to or greater than the average EU price. For pork and poultry, the baseline assumes that market prices adjust to clear the internal market.

The effect of the herd liquidation program because of the "mad cow" crisis is included. Continued limited intervention for beef, a shrinking dairy herd, and measures to encourage less intensive production methods are also assumed to limit beef production. To prevent surpluses from accumulating in the face of lower consumption, it is assumed that revisions to the CAP will further reduce beef producer incentives. The baseline assumes that there is no enlargement of the EU-15 to add one or more Central or East European countries. Accession of the large agricultural-producing Central and East European (CEE) countries could cause serious problems for the CAP in its current form and would likely require changes in that policy.

Asia and Oceania

Australia: Australia exports the majority of its crop and livestock output; this continues in the future. The Australian Wheat Board (AWB) is being reorganized so that it will be more along the lines of a commercial business with grower ownership and control. The AWB will retain single-desk status for exports, at least for the next 5 years. Australia is periodically subjected to drought, so adequate water availability is crucial to attaining the output levels projected. Crops are once again being planted in the Ord River project in Western Australia and several new dams are being planned, but it is still too early to factor in the full extent of the additional area for irrigated crops such as cotton and sugar. With the return of better weather after several years of drought, cattle herds are being rebuilt. Any dependence on imported feed during periods of shortages has been ruled out for the time being, which is dampening growth expectations for fed-beef. More favorable returns for other enterprises and low export prices are also limiting fed beef growth. Continued growth in exports of live cattle is projected, reducing the availability of suitable cattle for feedlots.

China: China's economy is assumed to continue to grow at a rapid, but declining, rate in the baseline. Average real GDP growth is forecast to slow from more than 10 percent in recent years to 8.4 percent in 2001 to 2005. This projection assumes China will continue its gradual reform of the remaining areas of the economy where there continues to be a substantial degree of government intervention and control.

Major reform initiatives will focus on the industrial sector and the political, social, and economic difficulties involved in reducing and restructuring the state-owned sector. Continued rapid growth in domestic and foreign investment allows the development of port, rail, road, and power generation infrastructure to keep pace with increased trade flows and energy demand.

Agricultural policy continues to move gradually and incrementally toward greater liberalization, increasing the role of market forces in China's production, consumption, prices, and trade. Central government planning is assumed to decline for most crops, with a growing share of farm gate and retail purchases occurring at market prices rather than at government-set prices. Intermittent government intervention to stabilize markets occurs, but with declining frequency.

Agricultural trade is assumed to continue its recent course, becoming more liberalized as tariffs are gradually reduced and non-state trade companies become more important. While central government control over trade in key commodities will likely continue, the share of trade handled by private or joint private-public trade companies is assumed to grow. In the baseline, it is assumed that China will not become a member of the World Trade Organization (WTO). China has applied for WTO membership, but negotiations are ongoing, with timing of possible entry and provisions of the final agreement uncertain. Production of all major crops (except rice) is projected to increase as rising domestic prices raise yields through increased use of improved varieties and fertilizer and better farm management practices.

Reduced state investment in agriculture during the 1980's produced a slowdown in the rate of yield growth toward the end of the baseline. Total land in agriculture will continue its current decline under pressure from nonagricultural uses, but the rate of decline will slow in response to a continuation of the government's more effective land management policies of the 1990's. Income growth that will drive demand for meats and edible oils will be the key factor in China's future agricultural trade patterns. Relatively small changes in assumptions affecting income growth, meat production trends, or feed demand can result in relatively large changes in trade projections for a country with 1.2 billion people.

East Asia: South Korea and Japan continue to open their livestock sectors to foreign competition as dictated in the Uruguay Round, using deficiency payments to assist the beef cattle sector and encouraging pork and poultry production with indirect subsidies. Japan will also make maximum use of the pork and beef safeguard mechanisms negotiated in the Uruguay Round, which raise tariffs and levies on those meats on a quarterly basis. South Korea, Japan, and Taiwan will retain bans on livestock imports from areas with foot-and-mouth disease.

All three East Asian economies are assumed to maintain tight state control over the trade in rice. Rice production in South Korea will continue to be insufficient to meet domestic needs, so it will have to import at levels above Uruguay Round mandates in some years to replenish stocks. Japan will continue to meet its minimum access commitment, but will not import above those levels. Rice imports of Japan and South Korea are projected to remain at the final levels set by the Uruguay Round for the years after 2000 and 2004, respectively. Japan's wheat, barley, and soybean production, and South Korea's barley and soybean production are maintained through border protection and the use of domestic products by processors in response to government mandates or subsidies.

The new quota for corn for new industrial uses should expand Japan's nonfeed market for corn. The East Asian governments will continue enormous expenditures designed to help domestic agriculture restructure itself. A continued steady outflow of labor from farming will help full- time farmers achieve larger operations and economies of size. Despite the restructuring, production of some key commodities will decline in some countries, including rice in South Korea and pork and poultry in Japan. In South Korea, declining rice consumption will mean that production declines may not lead to increased imports; in Japan, however, greater pork and poultry imports will be needed to offset the production decline. Southeast Asia.Rising incomes and the changing diets that follow are turning Southeast Asia into an expanding market for wheat and feedstuffs. This trend has been boosted as several countries have liberalized controls on imports of agricultural products. This trade liberalization trend will continue.

Improved economic conditions in Vietnam and Burma will help the region regain its former prominence as a net exporter of rice. Rice imports by several countries, including Indonesia, the Philippines, and Malaysia, will rise as policy goals gradually shift away from full selfsufficiency. The Southeast Asian region has abundant land resources well suited to rice production. Demand for wheat-based products will be increasing, but because the region's tropical climate is not suited to wheat production, rising demand will be supplied through wheat imports. With the region's economies continuing to boom, a key driver of agricultural import demand will be the expansion of livestock production, especially poultry. Most growth in livestock product demand will be met by local production that is increasingly dependent on imported feeds.

Thailand, the only significant corn exporter in the past, has recently become a net importer, with net imports projected to continue to rise. Low corn yields in the region limit corn's competitiveness with other crops, so corn production will not expand as rapidly as demand. The region's imports of soybeans and soybean meal will also show strong growth to meet feed demand, as soybeans are not generally well suited for production in the region's tropical climate.

South Asia: India's farm sector will continue to benefit from improving terms of trade as agricultural price incentives are maintained and liberalizing reforms steadily reduce protection in nonfarm sectors. Food grain production will also be given a boost by reduced protection of oilseeds resulting from the recent tariffication of vegetable oil imports. Domestic surpluses of rice continue in the baseline, with India's relatively low-quality rice maintaining a significant global market share. While some wheat exports are projected, India's surpluses of relatively low-quality wheat are more likely to be disposed of in the domestic market. With the reform of vegetable oil trade remaining in place, vegetable oil imports will grow rapidly. Price incentives and productivity gains will sustain strong growth in cotton production, with most production consumed domestically to meet domestic and export demand for cotton-based products. Producer incentives in Pakistan will continue to support gains in cotton acreage, leading to continued stagnation of wheat yields due to late planting. Trade policy permits rising dependence on imported wheat. Cotton yields will recover gradually from current pest-related problems. As with India, most cotton production is processed domestically, with strong growth in exports of cotton-based products. Continued, relatively liberal import policies will permit continued growth in vegetable oil imports. Growing livestock product demand may lead to emergence of significant corn and soybean meal imports.

Africa and the Middle East

Sub-Saharan Africa: Little or no growth is expected in per capita incomes and, with slow growth in production, constrained import capacity, and strong population growth, per capita food grain consumption is projected to continue to decline. Capacity to import food commercially will grow slowly, consistent with gains in total export earnings and real declines in food prices. The region is projected to receive a growing share of available global food aid. However, with global food aid budgets assumed to be fixed at current levels, food aid to the region will not be sufficient to maintain per capita consumption.

North Africa: Stronger growth in import demand for grains and feeds is projected in most of North Africa, based on the outlook for improved economic growth in most countries, limited production potential and, for some countries, more open trade policies. Political unrest will constrain economic growth in Algeria, but wheat and corn imports are projected to rise as crop production is hampered by high input prices, input shortages, and lack of credit. In Egypt, average real GDP growth of 4 to 5 percent annually and recent policy reforms will generate more growth in wheat and corn imports. Since joining the WTO in 1995, Egypt has been reducing producer and consumer subsidies in agriculture and has opened up trade to the private sector for some grains, cotton, and other commodities. Morocco's real GDP growth of about 5 percent annually, coupled with a continuation of recent steps to liberalize trade in grains, oilseeds, and sugar, will spark stronger growth in import demand. In Tunisia, which began liberalizing its domestic markets and trade in 1992, real GDP growth of 5 to 6 percent a year will lead to rising import demand for wheat and livestock products.

Middle East: Many Middle Eastern economies will also experience stronger economic growth during 1997-2005, in large part due to the outlook for rising oil prices. Prospects for Iran are highly dependent on both oil prices and the implementation of structural reform. Moderate economic growth, together with limited success in improving yields, and an ambitious livestock/dairy development program, lead to the projected growth in rice, corn, and barley imports. The situation in Iraq, both economic and political, is extremely uncertain. Under the assumption of 3.5 to 4 percent annual real GDP growth, food consumption is projected to gradually recover from the sharp drop following the Persian Gulf War in 1991, driving moderate growth in imports of food and feed grains. In Saudi Arabia, economic growth will also improve because of stronger oil prices, while agricultural output will continue to decline as budget constraints force the government to cut subsidies and there is rising concern about depleting water resources. Rising imports of rice, wheat, and feed grains are projected. Turkey's many difficulties, including high population growth, large external debt, and no strong commitment to privatization, will affect its economic performance well into the projection period. Steady growth in rice imports is likely, reduced producer subsidies are expected to push up wheat imports, and livestock and poultry development efforts should strengthen feed import demand.

Western Hemisphere

Canada: A major factor affecting baseline production projections for Canadian crops is the shift over the past several years into the production of canola. Encouraged by development of new varieties, canola acreage rose from a range of 2.5 to 3.7 million hectares during 1984-92, to a range of 5.3 to 5.75 million hectares during 1994-95. Canola plantings significantly affect area and production of other crops, particularly wheat and barley. Wheat acreage, for example, was below 11.3 million hectares in 1994 and 1995 after remaining well above 13 million hectares over the 1984-92 period. In 1996, prices strongly favored a return to grains, but the tendency to substitute canola for wheat acreage is projected to reemerge in the near future. However, rotational constraints on canola plantings are assumed to limit canola acreage.

Canada's 1995/96 and 1996/97 budgets projected a reduction in annual domestic support programs for agriculture from C$854 million to C$600 million over 3 years. In redesigning agricultural support programs to meet the new budget restrictions, emphasis is being placed on providing whole-farm insurance (such as the recently developed whole-farm savings plan program-the Net Income Stabilization Account), rather than crop-specific and production-distorting subsidies. The baseline assumes that government subsidies to crop and revenue insurance programs will be "production neutral" and that Canadian grains and oilseed production will fully respond to market forces. Canada's 1995/96 budget also eliminated the C$561 million Western Grain Transportation Act (WGTA) freight subsidy for prairie grains and oilseeds, effective August 1, 1995. The elimination of the WGTA freight subsidy meets Canada's commitment under the Uruguay Round export subsidy reduction requirements.

Elimination of the subsidy means that the cost of transportation of prairie province crops (such as wheat, barley, and canola) to export positions has increased by about C$17 per metric ton. This increase in transportation costs will reduce farmers' incentives to plant grains and oilseeds and will reduce production. At the same time, prairie processing and livestock sectors will benefit from reductions in local prices. The WGTA subsidy removal has reinforced recent trends toward more value-added processing in the Canadian prairie region. Substantial increases in livestock feeding and canola crushing are projected to continue in the baseline.

Increases in Canada's wheat exports to the United States over the 1990-94 period led to the negotiation of a bilateral agreement to govern wheat trade with a tariff-rate quota for 1 year, from September 12, 1994, to September 11, 1995. The agreement also established a joint commission to study all aspects of U.S. and Canadian grain marketing systems. With expiration of the tariff rate quota (TRQ) in September of 1995, USTR and USDA announced that the United States now planned to "monitor" imports of Canadian wheat using the expired TRQ as a benchmark for comparison, and to ask for consultations with the Canadian government if there is a surge in imports. The baseline assumes that these provisions will prove sufficient and that no new restrictions on U.S. grain imports from Canada will be imposed.

Several commodities grown in Canada have unique characteristics that are likely to guarantee certain export markets for the future. Canadian canola is preferred by Japanese importers. Canadian oats are an indispensable import for U.S. processors. Canadian and Australian barley malt are positioned to benefit from increasing demand from importers in China and Latin America. Because of these market "niches," projections for Canadian production of these three commodities is favored in the later years of the baseline.

*Mexico.*The economic crisis in Mexico triggered by the peso devaluation in December 1994 did not fundamentally change the long-term outlook for Mexican agriculture. The economy is expected to bounce back relatively quickly, with annual real GDP growth exceeding 4 percent in 1997 and averaging more than 5 percent through 2005. Mexico will be a progressively larger importer of grains, oilseed products, and meats over the next decade. Mexico's productive capacity will be limited by scarce water, land, and low levels of technology. Growing demand for meats will spur domestic meat production and demand for imported feed ingredients. Trade liberalization provides opportunities for greater imports of meats, almost entirely from the United States.

Agricultural policy continues to be driven by the PROCAMPO program and NAFTA. Under PROCAMPO, the government will continue to reduce its role in supporting grain prices. Intervention in domestic corn and wheat prices will end and, with lower import duties on corn, sorghum, and wheat, there will be more price transmission between the world and domestic grain markets. PROCAMPO direct payments, which require planting but are otherwise decoupled, will continue to be phased out. Under NAFTA, all tariffs on baseline commodities will be eliminated by 2008. In light of the price-competitiveness and quality of U.S. corn, pork, poultry, and eggs, particularly to the border areas, it is assumed that Mexico will import at least the quantity specified by the tariff-rate quota.

Mexico continues to reduce consumer subsidies, and the main subsidies that continue will be those on tortillas and milk. Feed compounders will now procure corn directly from farmers, thus eliminating CONASUPO subsidies for animal feed. South America.Strong overall economic growth is expected in South America, led by the two largest economies in the region, Brazil and Argentina. Many countries in the region will continue to benefit from their successful evolution from semi-authoritarian political systems and market-managed economies to political pluralism and market-oriented economies.

For Argentina, the key assumptions are on the supply side and involve the availability of land for crop production and the level of yields obtainable. Before 1996, the most area that Argentine producers had devoted to grains, oilseeds, and cotton was 18.3 million hectares in 1983. Until recently, it was considered unlikely that this record could be exceeded without significant investment in improving the marketing infrastructure. In 1996, however, planted area will surpass the record by more than 2 million hectares.

As a result, the baseline assumes that cropped area can expand significantly beyond 18.3 million hectares when market conditions provide adequate incentives. Crop yield response in 1996 also indicated stronger response to prices than in the past, with the use of inputs increasing sharply. Consequently, the baseline assumes faster growth in use of fertilizer and other inputs than has been the case historically. Finally, it is assumed that Argentina will not complete the process of attaining hoof-and-mouth-free status in the baseline, thus preventing expansion of beef exports to hoof-and-mouth free areas.

In Brazil, the economic stabilization program begun in July 1994 has used tight monetary policies to bring inflation down to about 9 percent, the lowest level in 39 years. The Central Bank will continue to manage a gradual devaluation in the real exchange rate in an effort to keep inflation and the growing trade deficit under control. With gradual real depreciation of the exchange rate, Brazilian producers will face stronger price incentives in local currency terms. Finally, in August 1996, the Brazilian Government eliminated the value-added state tax on exports of raw and semi-manufactured agricultural products. This change will have a significant positive impact on price incentives in the soybean sector, increasing the quantity of soybeans produced, as well as the volume of soybeans, meal, and oil that is exported.

Transition Economies

Former Soviet Union: Between 1997 and 2000, real GDP growth for the countries of the FSU will be very sluggish, and currencies will appreciate in real terms. After 2000, real GDP growth across the region is assumed to be 3 to 4 percent per year, with the exchange value of the region's currencies remaining roughly constant in real terms. The projections assume that liberalization of the markets and restructuring of agricultural enterprises of the FSU will continue at its current slow pace. Commodityspecific trade policies will remain mostly unchanged, with tariffs remaining at relatively low levels, and no quotas imposed. Price transmission between world and domestic markets for major commodities is assumed to be about 50 percent, meaning that a 1-percent change in the world price will result in about a 0.5-percent change in the domestic price.

The primary policy uncertainty in the outlook concerns the possibility of more protectionist trade measures for agricultural commodities. Higher tariffs and/or tariff-rate quotas or quotas may be announced in Russia for livestock products. Significantly higher tariffs, or imposition of quotas, could drastically change the meat import projections. Tariffs will probably be raised, but more drastic changes that could affect meat imports, such as quotas, will probably be avoided.

Crop productivity gains in the FSU will be small. Progress in land reform that could lead to major productivity gains is not anticipated. FSU livestock production is assumed to recover very slowly, at least until the process of economic reform reduces production costs and increases the competitiveness of the sector. The current high cost of meat production in the FSU suggests that livestock inventory declines of recent years will not be fully recouped in the foreseeable future and some meat demand will continue to be satisfied by imports. It is also anticipated that grain imports will be minimal in the baseline because continued livestock declines will limit demand. The Central Asian countries of the FSU will meet their grain needs primarily from Kazakhstan and Ukraine, rather than from imports from abroad.

*Central and Eastern Europe:*The baseline assumes that none of the CEE countries will join the EU during the projections period. Although some CEE countries may join the EU by 2005, the timing of accession is uncertain. The baseline also assumes that most world prices are fully transmitted to domestic markets and that import tariffs in most cases will not exceed 30 percent. In the short term, the impact of protectionist policies in the Visegrad Four (Poland, Hungary, the Czech Republic, and Slovakia) has mainly been to keep domestic producer prices at world levels. These measures have tended to counter the downward pressures on prices coming from the lingering bottlenecks in the downstream sectors. Of the Visegrad Four countries, only Hungary seeks to be a major grain exporter. Others aim for self-sufficiency. The baseline assumes that domestic producer prices will not differ greatly from world market prices. The principal constraint will be continued pressure to keep state budgets in balance.

The baseline also assumes a steady increase in efficiency in the agricultural sector, which will be reflected in rising yields and greater feeding efficiency in the livestock sector. These productivity increases will come about as a result of continuing progress toward market reform in all the CEE's. It is assumed that most of the rigidities inherited from the Communist period will be removed, allowing fuller transmission of world market prices to domestic producers. In addition, the forecast assumes continued positive income growth and falling inflation. Rising incomes and lower interest rates will bring badly needed investment to both agriculture and food processing. There will likely be some consolidation of the small fragmented farms that currently dominate much of the landscape. Land tenure will become more permanent, bottlenecks in issuing titles will be resolved, and true land markets will develop as capital markets improve.

7

Tools for Agricultural Research and Planning

Numerous planning tools are available to help planning teams understand the future environment of an organization and how it might respond to challenges and opportunities that may arise. Planning tools contribute to the planning process in two dimensions.

First, they support the planning process by laying out clear procedures and expected outputs and by defining rules for communication. The future is too multifaceted to be fully understood by a single person, and within planning teams different perspectives on the future may exist. Through the use of tools, different personal qualities in developing responses may be combined in a structured manner to arrive at balanced decisions with broader support than would be possible if single individuals were to undertake the planning effort.

Second, they enable planners to integrate external information in the planning process.

There are many sources of relevant information and many analytical perspectives on what the future may bring. Tools with which to process this information into a form that is readily understood by many people improve the quality of the resulting plan.

SUPPORTING THE PLANNING PROCESS

Bringing people and perspectives together in the planning process is one key responsibility of the planner or planning unit. Most of the people involved will not and should not be professional planners. Their central interest in the planning process is how it affects their work, their role, or their living conditions. In order to use participants' time as effectively as possible, tools are useful for structuring and organizing planning. These tools provide guidance in the planning task, help to define the objective of the planning process (or a step in it), and help people organize their thinking. Tools to support planning processes have often arisen from trial and error in the actual development of plans. Most, therefore reflect a certain judgment of what is or is not a good plan.

Sometimes a series of tools constitutes a complete planning procedure. For example, the German technical assistance organization, GTZ, developed a series of planning-by-objective procedures that they use in planning most of their projects. The procedures may be especially useful when the nature of the planning exercise is well known and agreed upon. If this is not the case, it may be difficult and undesirable to follow a standard procedure. The challenge then becomes choosing and combining tools to support the planning process in the best manner possible. It may even be impossible to fully anticipate which tools will be required.

Tools to support planning tend to concentrate on facilitating communication or on outlining the planning process. To facilitate communication, some simple tools may greatly contribute to the effectiveness of a planning group. Personal introductions before starting to work together are useful, especially if people talk about why they are part of the planning team and what they hope to contribute. Brainstorming is a well known approach to gather numerous ideas, which are organized afterwards. The "metaplan method" is among the best known tools for improving communication. It facilitates understanding and recollection of the discussion by the use of cards on which central thoughts are expressed.

The metaplan method is used in the constraint-tree approach. Tools such as the analytic hierarchy process have been developed to help decision makers evaluate alternative options. Scoring methods are similar but less formal. Decision support tools are most effective if they are based on a simple evaluation mechanism and lead to easy-to-understand results. Whereas these methods are effective for guiding the decision process, the risk is that they stress subjective over objective (evidence-based) knowledge. With the advent of information technology, computer-based tools for supporting planning have gained prominence. An advanced tool is the "decision room."

Here, a number of people use a decision support method such as analytical hierarchy process or scoring and evaluate alternatives, immediately discussing agreements and disagreements. Project planning and management software, such as Microsoft Project, may be used to assign responsibilities, fix schedules, and define financial requirements over time. "GANNT charts," which are used to map activities and time throughout execution of a plan, are commonly included. "CPM" (critical path management) or "PERT" may also be used. These tools are used to optimize planning activities in highly complex projects so as to be completed in a minimum of time. Where such software appears too complex, responsibility charts may be useful. A simple responsibility chart might indicate the task, persons involved, and the time when a planned activity should be undertaken. Computer software planning tools tend to be more appropriate for planning activities with a short time horizon and a limited scope (e.g., project planning), rather than for planning with a long time horizon and a wide scope (e.g., strategic planning).

Another set of planning tools serves to integrate more and higher quality information in the planning process. These tools can improve planning in at least three ways:

1. They may help to predict the future environment of the organization and therefore allow for plans that better anticipate changes that may occur.
2. They may improve predictions of expected impacts of possible plans, thereby improving the evaluation of alternative plans.
3. They may improve estimates of resource requirements and resource availability, thereby increasing plan feasibility.

Tools that improve the information base of the planning process have often originated in a specific scientific discipline. For example, cost-benefit analysis, grounded in economics, contributes to understanding expected impacts of alternate plans in the light of resource requirements.

Geographic information systems (GIS), from geography, help planners understand the spatial impacts of new technologies. The most useful tools for improving the information base are those that are easily understood by all people involved in the planning exercise. The maps that can be produced using GIS are easy for most people to understand and interpret. In this respect GIS is an excellent example of a planning tool that produces highly accessible outcomes. There is an almost infinite number of tools that can be used to improve the information base in planning. Planning, however, may suffer from too much as well as too little data. If information to improve planning comes from many different sources, reflects many kinds of parameters, or is based on a wide range of methodologies, the planning group may be overwhelmed and lost. One discipline may then start to dominate the planning process. Tools to improve the information base should focus on the most essential dimensions of the decision problem, or they should allow participants to truly combine more information towards a relevant plan.

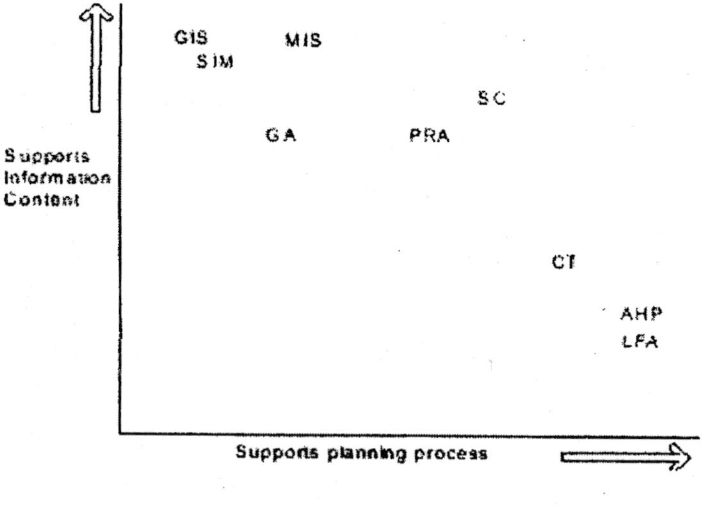

Figure 1. Contributions of planning tools to the information content and planning process

Although many tools contribute to both the dimensions referred to earlier, normally they contribute more to one than to the other. Figure 1 positions the nine tools described in the following chapters on these two dimensions. It suggests that more process-oriented tools might be best combined with tools that are more information based.

As explained, these are certainly not the only tools that can be usefully applied in a planning exercise. Other tools may be introduced depending on the nature of the planning process. Using the analytic hierarchy process (AHP), complex decisions that are difficult to oversee are broken into smaller parts that are more easily handled. A general goal for the decision-making exercise is defined, and the goal is decomposed into a number of criteria. An assessment is made as to how alternative research programs or projects contribute to the different criteria. This is done by comparing pairs of alternatives with regard to each criterion. The assessments on each criterion are then combined into an overall assessment. Constraint trees are used to develop a shared understanding of the problems and opportunities that applied research must address in a specific domain, such as a commodity or region. It is an excellent tool for guiding the diagnostic phase of the planning process, because it allows many different people to participate and many ideas to be reconciled. By distinguishing the different problems and opportunities faced by men and women, gender analysis allows research programs to target their interventions with more precision to users. Gender analysis is typically used to strengthen the diagnostic phase of planning. It is most useful at the research program and project levels.

Geographic information systems (GIS) help planning participants understand the spatial pattern of problems and opportunities in agriculture and to predict the spatial impact of research. This is useful for evaluating the expected success and impact of research strategies. GIS enables results to be presented in attractive maps that are easy for most people to interpret.

Logical framework analysis has become the standard for summarizing a project or program plan. The logical framework describes the causality relating activities to outputs, objectives, and the final goal of a project. It also identifies mechanisms for verifying progress towards outputs and objectives, as well as the critical assumptions that must fulfilled for the project to be successful.

A management information system (MIS) includes many elements outside of planning. But it also supports planning processes by providing information on resource use and availability and by feeding results of planning exercises back into the information system for monitoring. Participatory rural appraisal (PRA)may strengthen the information base for planning as well as affect planning procedures. Within the realm of PRA there are many tools for collecting information from farmers and other rural inhabitants.

In PRA both information extraction and deciding how to use the information are participatory. PRA thus brings research planning closer to the user. Scenario analysis helps planning participants view the future without prejudices that may arise from past experiences. If there is considerable uncertainty about the future and the implications of alternative decisions are very significant, scenario analysis may help improve decision making. Finally, simulation models are used to evaluate the impact of alternative plans on the research subject. By using all the information available on a subject and complementing this with additional experiments, models can be constructed that allow prediction of behavior under a series of assumed conditions. Especially for complex subjects simulation models may clarify the attractiveness of alternative research strategies and may help planners choose the optimum one.

LEGAL ISSUES IN AGRICULTURAL RESOURCE MANGEMENT

The field of legal issues in Geographic Information Systems (GIS) is evolving rapidly, and various international agreements and national laws are under discussion that will affect use of spatial data and software. Key issues relate to respecting the rights of owners while avoiding barriers to the flow of data and processed information needed for research in agriculture and natural resource management.

Copyright Issues

The term "copyright"originally referred literally to a "right to copy" the literary works of an author, and was meant to encourage the publication of written works. Copyright was implemented by requiring publishers to secure permission from authors to reprint their work. Copyright provides a public benefit with little burden on citizens, and worked reasonably until advances in digital reproduction began permitting very low cost, high quality copying. Rights holders thus are taking increasingly drastic steps to protect their products, and some of these measures arguably impinge on the long-established rights, such as fair use, of certain users.

The definition of "works" has expanded significantly over past decades, as new technologies permit ever more diverse and powerful ways of expressing ideas. Such works must normally be original and creative in nature. A mere recitation of facts, regardless of how expressed, does not automatically qualify for copyright. Under Article 4 of the

WIPO Copyright Treaty..1996 and by Article 10 of TRIPS, 1995, computer programs are considered literary works.

Copyright grants certain protective rights to the owners of the expression of an original idea, formula, model, method, or theory. Contrary to patents, which protect concepts or ideas that have been developed to a practical level, copyright provides no protection for ideas per se, only for the form of expression.

The duration of copyright protection varies by country. The Berne Convention (Article 6) and TRIPS (Article 12) provide for protection for the life of the author plus 50 years as the minimum, but many countries specify a term equal to the lifetime of the author plus 70 years. For works with multiple authors, the term of protection is measured from the death of the last surviving author (Article 7). Under the US Copyright Act of 1976, anonymous works, pseudonymous works, and "works made for hire" are protected for 75 years from first publication or 100 years from the year of creation, whichever expires first.

Two important protective rights granted to the copyright owner are control of reproduction, whicn prohibits copying, and control of translation. A translation acquires its own copyright, but this cannot infringe on the owner's prior rights; that is, the translation must be authorized by the owner and must acknowledge the owner's copyright in the original work. The copyright owner does not usually need to formally register the material to be protected. Formal registration may be important for legal action in claims of copyright infringement. The details of enforcement of rights can also affect an owner's ability to assert his rights.

Most countries no longer require that a copyright notice be displayed. Older works that do carry the (c) notice may be in the public domain, due to expiration of protection, and some works may claim protection but, in fact, do not warrant protection. One example of the latter case is a work containing simple lists of facts, perhaps in tables, which a court determines to have neither originality nor creativity (nor sweat of the brow, in some jurisdictions) in compilation. Copyright infringement occurs in three forms:

1. Deliberate infringement occurs if the defendant used unauthorized copies in a manner that violates fair use.
2. Contributory infringement occurs when the defendant knowingly enables another person to infringe on a copyright.
3. Vicarious infringement implies that, although the defendant did not engage directly in infringement or enable another to infringe, the defendant had the right and ability to supervise the infringing activity and had a direct financial interest in the infringement.

Penalties for infringement under US law can run as high as US$100,000. Infringement proceedings can be taken against researchers and their institutions. Risk of vicarious infringement increases in projects that involve multiple organizations operating under different legal jurisdictions and IP regulations or cultures. As more works become digital and original works are transcribed to digital formats, they can be published-that is, copied and transmitted-at minimal cost and without knowledge or control of the author or owner of the work. This ease of abuse has undermined the basic principles of copyright and enforcement, leading to calls for stronger legal protection. The US Digital Millennium Copyright Act (DMCA) prohibits many actions not considered infringements in other laws governing copyright. Citizens in other countries face similar inconsistencies.

Copyright and Spatial Information

Researchers creating or using spatial information need to understand the basic principles of copyright protection, including possible consequences of infringements and specific issues concerning protection of databases. In many projects, more than one copyright owner will provide data. An agricultural researcher may access a soils database copyrighted by a private company, whereas the land (parcel) boundaries used in the research may come from a government agency. To collect data relating to plant growth and to analyze and present the results, both sets of data are required. One source may willingly permit data to be used freely or at low cost, or the data may be in the public domain. The other data may be privately held or under control of a government agency acting under a strong cost recovery regime.

Fair use provisions may apply, depending on the amount of data used by the researcher, or special exemptions for non-commercial use by researchers or educators may apply. Provisions may apply under the national law of one data provider but not under that of another, complicating crossborder data interchanges. Such issues are best resolved through a careful inventory of data sources and associated IP. Detailed guidelines on how to inventory data sources and associated IP are available at the CAS web site. Digital databases are not explicitly protected by copyright under the Berne Convention. The WIPO Copyright Treaty, 1996, extends copyright to "...compilations of data (databases)" as "intellectual creations;" but this protection may not cover the data itself, as facts are not viewed as the result of a creative process.

Copyright provisions in the WTO TRIPS Agreement (Article 10.2) give some measure of protection to databases. A proposed WIPO Database Protection Treaty would extend copyright protection to databases.

Copyright and Fair Use and Fair Practice

Copyright regulations provide for various public uses of works. The TRIPS Agreement and Article 10 (Certain Free Uses of Works) of the Berne Convention define these exemptions. Article 10 outlines fair practice in using several types of works, including access to databases:

1. It is permissible to quote from a work available to the public, provided that the quotation is compatible with fair practice and the extent of material used does not exceed that justified by the purpose.
2. It is up to national legislation or agreements between countries subscribing to the Berne Convention to permit works to be used to the extent warranted "by way of illustration in publications, broadcasts or sound or visual recordings for teaching, provided such utilization is compatible with fair practice."
3. When protected works are used in accordance with Article 10, the source and the name of the author should be published with the quoted material.

The fair use doctrine thus moderates a copyright's capacity to suppress expression. Unfortunately, there are no absolute tests to identify fair use. US courts issue opinions on individual cases, and it is unadvisable to extrapolate from one case to the next. Eisenschitz and Turner summarized the main copyright exceptions for fair dealing embodied in national law in the European Union, as follows:

— Copying for personal use.
— Copying for scientific, educational or other private use.
— Archival copying and library privileges.
— Educational exemptions and (other) graphic reproduction.

For alleged illegal use of copyrighted material, the main defense is under fair use exemptions. Applying fair use can be especially difficult though in the case of a database of facts.

Legal Protection of Databases

Databases used by scientists may be in the public domain, copyrighted as compilations under the Berne Convention, protected by database laws, or protected as trade secrets and accessed through licenses. There is much debate on appropriate means to protect databases, due to the wish to reward large investments in compiling facts vs. the recognition that simple compilations are neither copyrightable nor patentable.

Copyright Protection for Databases

Article 5 of the WIPO Copyright Treaty, 1996, and Article 10 of the WTO TRIPS Agreement, 1995, extend protection to "compilations of data (database)," i.e., "compilations of data or other material, whether in machine readable or other form, which by reason of the selection or arrangement of their contents constitute intellectual creations shall be protected as such. Such protection, which shall not extend to the data or material itself, shall be without prejudice to any copyright subsisting in the data or material itself." These two treaties formally extend the definition of protected works contained in Article 2 of the Berne Convention. However, the definition of a compilation remains similar to that in the Berne Convention for literary works and does not approach the definition of a database set out in the EU Protection of Databases Directive. Linne states that: "By law, the US federal government cannot copyright databases, although private vendors disseminating government information can. Scientists can generally use copyrighted material because of a fair use exception in the United States or equivalent exemptions in Europe."

EU Legal Protection of Databases Directive

To protect investments in large databases and harmonize disparate national legislation in Europe, the European Union adopted a directive providing a novel form of legal protection for databases in 1996. The directive has now been translated into national laws or transposed into existing legislation in the EU countries. Because of its scope and emphasis on rights of investors, this activity is potentially the single most important legislative development affecting users of spatial information in recent years. Features of the directive include the following:

— Protection is not based on extensions of copyrights or patents. It represents a unique or sui generis right (meaning "of its own kind").

— It protects "databases in any form," where database is defined as "a collection of independent works, data or other materials arranged in a systematic or methodical way and individually accessible by electronic or other means."

— The protection is "...for the maker of a database which shows that there has been qualitatively and/or quantitatively a substantial investment in either the obtaining, verification or presentation of the contents to prevent extraction and/or reutilization of the whole or of a substantial part, evaluated qualitatively and/or quantitatively, of the contents of that database."

— Extraction is defined as "...the permanent or temporary transfer of all or a substantial part of the contents of a database to another medium by any means or in any form" and re-utilization means "...any form of making available to the public all or a substantial part of the contents of a database by the distribution of copies, by renting, by on-line or other forms of transmission."

— In place of a fair use concept, the directive prohibits "...repeated and systematic extraction and/or re-utilization of insubstantial parts of ... the database ... which conflict with normal exploitation of that database or which unreasonably prejudice the legitimate interests of the maker of the database."

— Users may extract or re-utilize a substantial part of the contents of a database if this is done for private purposes or for "...purposes of illustration for teaching or scientific research" and as long as the source is indicated and the extent of copying is justified by the non-commercial purpose to be achieved.

— The term of protection for a database under the sui generis right runs for 15 years. Any substantial changes to the contents of a database, such as additions, deletions, or alterations that require a large investment, can qualify the modified database for a new term of protection.

— Databases that "...by reason of the selection or arrangement of their contents, constitute the author's

> own intellectual creation" are still protected by copyright. No other criteria need be applied to determine eligibility for protection. But copyright protection does not extend to the contents of such databases and is "...without prejudice to any rights subsisting in those contents themselves."

A reciprocity clause in the directive states that only countries that offer similar protection to EU nationals will receive this new level of protection within the European Economic Area. Many parties would agree that copyrights and patents provide little or no protection for the contents of databases and that investments by groups that assemble large databases merit some form of protection, thus regulating largescale copying and redistribution of the data. However, the EU directive has been severely criticized on multiple fronts, an underlying theme being that the protections excessively favor large commercial database providers at the expense of research and educational communities. One of the foremost concerns relates to the creation of the novel right. Copyright and patents have stood the test of time, and it seems risky to introduce a whole new class of protection without much more extensive consultation and deliberation. The directive also diverges from previous practice in automatically assigning rights to the investors rather than the creators. Critics further suggest that the provision for new periods of protection following database updates essentially authorizes perpetual protection, contrasting with the fixed time limits for copyrights and patents.

A fourth concern is that whereas terms like "fair practice" have resulted in extensive litigation in relation to copyrights, the directive contains a large set of descriptors such as "substantial part," "non-commercial purpose," and "substantial investment," that will have to be interpreted through court cases. Partly in response to the EU directive, WIPO proposed a Draft Treaty on Intellectual Property in Respect of Databases in 1996.

The draft generated strong disagreement among various sections of the information community, and this impasse has yet to be resolved. Because a large body of international research is done by or through institutions based in the EU member states, researchers should understand that the sui generis right is now active. Many spatial databases may have such protection, regardless of whether the immediate provider is aware of its scope and potential implications.

Protecting Databases by Other Means

Legal means such as contracts and technical measures such as encryption can also protect databases. Legal contracts have been used since the first on-line electronic databases were created for medical information, patents, and other scientific data. Contracts are used to control access to many large, spatial information databases, such as those created by many national mapping agencies and by satellite imagery vendors. Contracts have drawbacks, including "(1) a high administrative burden of negotiating terms with each user and provider of data, particularly for databases compiled from several sources, and (2) they cannot prevent unauthorized downstream uses of the database because they are only binding on the parties to the agreement." Whereas the first concern is being addressed through cleverly implemented, on-line, "click to use" licenses, the second concern is much more difficult to eliminate.

Technological methods such as encryption can also regulate use of databases, but they are expensive and complex to implement, both for vendors and users. The WIPO's Digital Agenda focuses on e-commerce in the information area and IP in particular, and addresses use of such methods. In the USA, the Digital Millennium Copyright Act (DMCA) enacts the provisions of the 1996 WIPO Copyright Treaty (Article 12) that prevents "...removal of electronic rights management information without authorization" from any protected work.

The DMCA further makes it an offence, with severe punishments, to publicize information that could lead to removal of protection mechanisms. Although not providing protection per se, data in large databases are often modified in a readily detectable way to permit reliable identification of the source, even when copied to new media or database software. Such digital watermarks may include manipulation of non-significant digits to create readily identifiable patterns and inclusion of records with fictitious data .

Patents

Contrasting with the focus of copyrights on forms of expression, patents deal with useful inventions. Patents are regulated under terms of the Paris Convention for the Protection of Industrial Property, 1883 - 1979 and Section 5 of the TRIPS Agreement. The TRIPS Agreement sets a minimum protection period for patents of 20 years "from the filing date" (Article 33). Patent protection must be requested from designated bodies in each country, following strict procedures. Applications usually are filed via patent agents who know these procedures. Agents also advise whether the application is likely to succeed, following a search for published information that might negate claims to novelty or non-obviousness.

Treaties such as the Patent Cooperation Treaty (PCT) and the Patent Law Treaty (PLT) simplify the application, search and examination processes. National applications must still be filed with the patent office of each nation, except where regional systems exist, such as the Organisation de la Proprie Intellectuelle (OAPI) in the francophone countries of Africa. The filing cost ranges from a few hundred US dollars to more than US$ 10,000. Maintenance fees can further increase the cost and, if necessary, defending a patent can cost millions of US dollars.

Since patent protection is only a right to prosecute infringement, the potential costs should be taken into consideration when deciding to file for a patent, and if so, where to file. Patent protection is sought not just for economic benefits and to restrict use of the patented elements, but to secure IP control of important new techniques, models, or tools which can then be released to the public, including other researchers, in a more managed way. Patents can protect the conceptual framework underlying intellectual property that has been rendered useful by a "reduction to practice." The danger to research institutions is that an organization will be granted a patent for a critical element of research, and this patent thus allows access only to those who can afford to pay the license fees. Patents are awarded after search and examination by experts employed in or contracted by national or regional patent granting authorities.

The validity of a patent depends on the skill of the examiner, the honesty of the inventor(s), and drafting skills in preparing the application. In recent years, there has been an outcry against patents that would have affected millions of users of information and communications technology equipment or services. Some were later overturned on request for re-examination, appeal, or in civil cases regarding fulfillment of patenting requirements. Some failed the novelty or usefulness tests. Others were disallowed on technical grounds when additional prior art was discovered or the innovation had actually been used, sold, or disclosed to the public prior to the filing of an application.

If a patent is infringed, the only remedy is generally a civil or administrative lawsuit. This contrasts with infringement of copyright, which can carry criminal penalties as well, and breach of confidentiality and unlicensed use of a trade secret without permission, which are criminal acts in most jurisdictions.

For agricultural research staff and their institutions, especially as regards spatial information, one key question is whether software tools, methods, and models are patentable in their legal jurisdictions. If so, then the possibility exists that third party products used in research are patented. In addition, the option exists to seek patent protection for research products developed by these staff or institutions.

Patent Protection for Methods and Processes

The patent system is designed to protect invention of both products and processes. Patent systems in some nations, such as the USA, grant patents for business methods, a practice not allowed in systems such as that of the European Union. The current debate centers on two issues:

(1) whether such patents should be allowed and, if allowed,

(2) how to better regulate the process by which they are granted, since patents are awarded for many seemingly obvious methods.

The US government supports the concept of patenting business methods, and a court ruling in 1998 found that "...such patents express the practical application (useful, concrete, and tangible result) of technology that is the essence of innovation."

During the consultative process preceding current attempts to create an EU Community Patent, the 19 countries that operate the European Patent Office (EPO), which examines European patent applications for a number of European countries, voted to allow patents "...in all fields of technology," including methods and software. According to some sources at the EPO, both software and some business methods claims have already been allowed by the EPO. In February 2002, the European Commission published a draft directive for patentability of software. The spirit of the document has been broadly attacked, and the fate of this initiative is as yet unclear.

Examples of Patents Related to Spatial Analysis

A search of the US Patent and Trademark Office's on-line database for "GIS" reveals the prevalence of GIS and related tools in the patent literature. The numbers refer to a drawing provided with the patent application. The patent is held by Multipmap.com in the UK, which reportedly has attempted to enforce the similar UK patent. Various data compression algorithms are patented. The GIF image format uses LZW compression, which is covered by patents held by Unisys and IBM. Software that allows manipulation of GIFs should include a license from Unisys.

Open source software specifically avoids use of GIFs to eliminate the need for licensing. Other patented compression algorithms include MrSID (LizardTech Inc.), which is based on technology licensed from Los Alamos

National Laboratory, and Enhanced Compression Wavelet 2.0 (Earth Resource Mapping, Inc.). These two algorithms are currently the subject of a patent infringement case. IDELIX Software Inc.'s Pliable Display Technology (PDT) extends display capabilities of traditional GIS software by providing "magnification-in-context" tools. As the user moves a "lens" over a display the image is magnified. Patents for PDT are pending. The patent covers portable devices that combine GPS and radio technologies to enable such devices to transmit positions to other such devices, as well as a method for indicating the location of one device on the display of another portable device.

Other Legal Issues

Other legal issues affecting use and misuse of spatial information include:

— Protecting commercial or organizational confidentiality.

— Ensuring data privacy (personal confidentiality) for individuals or groups.

— Abiding by national laws on exporting or importing data or information products.

— Understanding and incorporating appropriate licensing terms.

— Managing liability related to data, information and software products.

Consider the range of national and/or institutional regulations that might affect collecting spatial information from multiple data owners, analyzing the data using models or other input from other researchers, or the publication and distribution of maps created through a GIS.

Protecting Confidentiality in the Information Age

One method of protecting IP is simply to keep private as much information as possible about the product, process, technique, model, or method that is to be protected. Prior to the advent of patent regimes, secrets were one of the few protections available to inventors. Confidentiality. Confidentiality of information does not arise automatically.

A communication, in whatever form, should be clearly identified as "in confidence." Protecting confidences is governed by legal provisions differing from those for ideas. Common law and equity protect secrets via legal actions for breach of confidence. Laws in most countries consider it to be a breach of confidence if a person receiving information "in confidence" discloses the information to a third party. Almost anything, ranging from industrial formulas to lists of customers, can be protected. The main requirements are that the information must be "...secret, a discrete entity and uniquely ascertainable". The information does not need to be expressed in material form; verbal information is protected as fully as written information. The protection of confidential or trade secret information has become more standardized and enforceable internationally as a result of TRIPS provisions covering this area.

To maintain the confidentiality of electronic communications, include a simple confidentiality statement as part of the prefix or suffix to a message.

Research staff working on projects involving cross-border partners should be aware of any confidentiality clauses that may exist in project documents or funding arrangements relating to entire programs. If a research group involves institutions and research staff from different countries or different legal jurisdictions, contractual terms should be introduced into project agreements that can be used later to establish breach of confidence, should confidential information be knowingly or inadvertently misused.

Trade secrets are typically protected by unfair competition law. Protecting trade secrets in today's digital network environments relies heavily on technological measures such as encryption and password control. Once a trade secret has been stolen and posted to the Internet, courts may have trouble determining the "secrecy" element of the trade secret, as set out in TRIPS Article 39 (2-a and 2-c) below: Natural and legal persons shall have the possibility of preventing information lawfully within their control from being disclosed to, acquired by, or used by others without their consent in a manner contrary to honest commercial practices so long as such information:

(a) is secret in the sense that it is not ... generally known among or readily accessible to persons within the circles that normally deal with the kind of information in question;

(b) has commercial value because it is secret; and

(c) has been subject to reasonable steps under the circumstances, by the person lawfully in control of the information, to keep it secret.

Staff involved in agricultural research that generates information of commercial value should be aware of their responsibilities in regard to trade secrets. Such information is considered to be a trade secret if (1) the owner takes reasonable measures to keep the information secret and (2) the information has real or potential economic value by not being made public. Thus, not all confidential information is necessarily a trade secret, yet both types of information can be protected. A particular problem may arise when staff leave one organization and join another, taking both confidential and trade secret information with them simply because they were party to its creation or used such information in their work. Both staff and institutions should recognize their rights and responsibilities in such cases, and stated policies should be in place, including appropriate non-disclosure and confidentiality agreements and forms, both in contracts of employment and perhaps even on a project basis.

Personal Privacy

People working in agricultural and natural resource management may consider that such work normally does not invade anyone's privacy, yet seemingly innocuous records of farm expenditures, farming practices, or property boundaries are legally viewed as personal data, especially under recent European privacy laws. Countries with personal data legislation generally require that administrative steps be taken to inform the person(s) affected as to what is being collected, why, how, and for what use, as well as to register databases holding the personal data. Failure to follow these protective steps can lead to criminal prosecution, as under the UK's Data Protection Act 1995.

To avoid the need for such controls, a valid alternative is to eliminate the use of actual identities of any persons who might otherwise be identified via the research. The US National Institute of Statistical Science is developing a system for the US National Agricultural Statistics Service to disseminate geographical survey data on agricultural-chemicals usage that protects the identity of farmers. The information privacy policy promoted by the US Federal Geographic Data Committee to protect personal information in "geospatial databases" is as follows:

— When personal information is collected directly, at the time of collection agencies should inform individuals of:

  — why they are collecting the information and their legal authority to do so,

  — how the information will be used and protected as to confidentiality, integrity and quality,

  — the consequences of providing or withholding the requested information,

  — how to correct personal information if it lacks sufficient quality to ensure fairness in its use,

  — the opportunity to remain anonymous when appropriate and any rights of redress, plus the records retention schedule of the agency.

Personal information is to be acquired and used only in ways that respect an individual's privacy and such information will be collected only as needed to support current or planned activities. Agency staff should be are aware of the privacy implications of geographic information system technology.

Technical and managerial controls will be used to protect the confidentiality and integrity of personal information, including prevention of alteration or destruction of such information held in or linked to geospatial databases, and such information should be as accurate, timely, complete and relevant as possible for the purposes for which it is acquired and used. These policies apply to any US agency that collects personal information. They serve as an excellent model for policies to protect privacy.

Licenses and Licensing

Given the uncertainties over coverage by copyright and other protections, licenses remain one of the most common mechanisms used to protect software and data products. Research staff should understand the various types of licenses and the terminology used, both as end-users and as potential producers.

It is difficult to identify software licenses for commercial software packages that consist of less than several pages of small print. The ESRI Master License Agreement (MLA) has 11 major articles with numerous sub-sections plus a table and explanatory notes, as outlined below:

1. Definitions.
2. Intellectual property rights and reservation of ownership.
3. Grant of license (covering normal licenses, beta release licenses and evaluation licenses).
4. Scope of use (permitted uses and uses not permitted).
5. Maintenance.
6. Term and termination.
7. Limited warranties and disclaimers (3 subclauses).
8. Exclusive remedy and limitation of liability (2 sub-clauses).
9. Infringement indemnity.
10. General provisions.
11. Entire agreement, amendments.

A separate ESRI Data License Agreement has 15 clauses, including permitted and not permitted uses, redistribution rights for derived data sets, no warranty given for quality, limitation on liability, and export restrictions. Licenses for software or data downloaded from a Web site can be read by the user prior to accessing the information product. For packaged products, the paper version of a license is typically contained inside the physical packaging that contains the software or data. By opening the software package, most vendors lead users to believe that they have already agreed to the terms contained therein. Such "shrink wrap licensing" claims have been contested in various countries with varying results.

Considerable time and money may be spent in fighting claims by vendors against the purchaser in regard to defending the latter's right to refuse and/or return the software once the packaging has been removed. In the USA, courts have variously upheld and rejected the validity of shrink wrap or "click to agree" (on-line) license terms, in interpretations of the Uniform Commercial Code (UCC) that embodies US commercial law. A key issue is at what time did the purchaser have the opportunity to actually see the terms to which he/she had allegedly agreed, in relation to the time at which the product was purchased or first used. Researchers who offer their data under any conditions other than public domain need to carefully consider why the data are being released and what licensing terms are needed to achieve those aims. If data from multiple sources are integrated into the final result, careful consideration must be given to the licensing terms for each product. Unfortunately, few purchasers of software or data packs read licenses thoroughly and thus understand the legal conditions governing use of the product. Furthermore, the license text often requires an understanding of legal terminology to fully comprehend the ramifications of the agreement.

Two important topics used in licenses are those dealing with "uses not permitted" and "redistribution rights for derived data sets". Most such agreements prohibit transfer of data to unlicensed third parties. More importantly, since data acquired from outside a research project are likely to be integrated with data collected during the research, limitations on distribution rights for the "derived data sets" can be crucial. Such rights should be closely scrutinized at the start, to ensure that there is such a right, even if at an additional cost.

For data or tools made available to educational institutions, a statement is often provided that the product can only be used or disseminated for "noncommercial use." This concept also appears in the licensing agreements for many publicly available (but not public domain) spatial datasets, such as the 1:1 million scale Global Map project and for national datasets from many countries around the globe that are participating in the Global Spatial Data Infrastructure (GSDI) initiative. An educational institution using a topographic base map for classroom instruction has nothing to fear from such a restriction. A scientist who may have acquired similar data and uses them in completing a governing noncommercial commercial research project faces quite a different problem, including potential infringement.

Therefore, while apparently gaining ready access to no-cost or low-cost spatial information that is crucial to either their research (for spatial analysis purposes) or to the publication of final results, researchers should be aware of the terms and conditions of any external data sources used, unless the data are officially in the public domain. They should also beware of potentially losing rights to software or data that already purchased, as a result of upgrading to newer versions offered by the same supplier. Some licenses, covering trade-in upgrades, restrict the use of previous versions. However, most software upgrades and associated licenses impose no such restriction on the user or original purchaser, although all original terms of acquisition do still apply.

8

Trends of Agricultural Development

A Theoretical Framework for Agricultural Development

In Asian countries which have the potential for achieving self-sufficiency in staple foods, agricultural development is in a period of transition. Agricultural development in Asia since the 1980s has been characterized by the diversification of agriculture, and structural adjustment policies have been undertaken in many Asian countries during this period under the guidance of the World Bank and International Monetary Fund.

The diversification of agriculture is an inherent part of development strategy affected by structural adjustment and reflects the changing phase of development of the general economy. The optimum choice of crops for proper land use, the increase of agricultural production in parallel with the increase in demand, the production of agricultural products with higher value added, the processing of agricultural products, and their export promotion are some of the policy measures undertaken in the diversification of agriculture.

All these measures prioritize the principle of market efficiency which is a basic idea in structural adjustment policy. The increasing concern about the market has also generated momentum to reform agriculture which has been substantially subsidized by the government budget. Thus, interest in the actual state of the market itself in each Asian country has increased. In this context, the private sector has drawn increasing attention because it will replace the government which has led agricultural development. The agribusiness and agro-industry are expected to contribute substantially to agricultural development. In order to support numerous agricultural activities, a large number of local traders is crucial as a primary agent for the establishment and function of the market. The aim of this special issue is to analyze the institutional arrangements within the traditional market system which is relevant to agricultural development in rural Asia.

In Asian agrarian societies, economic transactions have been undertaken through the market, but these have not necessarily been well developed and organized. With the emergence of new goods or services as economic development progresses, a mechanism for their transaction needs to be created. To support their transaction, various kinds of institutions of transaction need to be created and consolidated because markets are not prepared in advance for the introduction of new goods and services. Institutions may not always function well; however, traditional institutions which have set down the rules of transaction can be invoked and adapted to carry out the transaction of new goods and services. Institutions and their arrangements are crucial in making a market function well.

When two individuals become engaged in a transaction of certain mutually agreed-upon goods or services, information on the transaction (especially a price or parameters which substitute or supplement the price) is formed, and this information somehow becomes available to a third party who wish to enter the transaction. If the entrance to and withdrawal from the transaction are potentially possible for all three parties, the relationship through which such a transaction can take place may be defined as "the market" in the broad sense. In traditional Asian villages, such economic transactions of various goods and services take place every day. In order to understand the market as defined above, we must suppose a complete set of perfect markets in the Arrow-Debreu sense. Here "complete set" means that there is a market for every commodity and service which the sellers and buyers want to transact, including the transaction of state-contingent claims. A perfect market has no transaction cost and no asymmetry of information. In the perfect market, each commodity has a price; competitive equilibrium is achieved; a rational incentive system is established; and thus resource allocation is efficient.

We define a situation which does not satisfy the conditions of the "complete set of perfect markets" (due either to an imperfect market or a missing market) as market incompleteness (or incomplete market). The real market that we simply refer to as "the market" implies this sort of incomplete market, namely, the "incomplete set of imperfect markets." The reality of the market in Asian villages is far from the ideal complete set of the perfect markets, particularly in the factor markets. In the rental market of land used for share cropping for example, the share of harvest crops as a contract parameter is fixed, and its function to meet demand and supply is insufficient. Interest rates in rural financial markets face the same problem. These are instances of an imperfect market. Insurance services as such are not traded in the majority of Asian villages. Therefore farmers themselves earn insurance effects through the transaction of other goods and services. Furthermore, many goods and services are invented as part of the process of development, or they become available through imports from overseas. Nevertheless, markets are not necessarily ready for these goods and services. These are instances of a missing market.

INSTITUTIONAL ARRANGEMENT

In the field of neoclassical economics, the completeness of markets has been regarded as a prerequisite for economic activity, therefore institutions have not been considered to have any influence on economic performance. For this reason, the study of institutions has not gained sufficient recognition as a research subject. Nevertheless, we consider that institutions are what complement or supplement the incomplete market. The developing field of the economics of information and the new institutional economics have enabled us to elucidate the economic transactions in such an incomplete market. When there is market incompleteness, a signal other than the market price (or a combination of price and another signal) will be used in the transaction. This is because the price in an imperfect market does not accurately reflect the economic value of goods and services, and the market price itself does not exist under a missing market. Long-term contracts are adopted, interlinkages among plural production factors or products are mobilized, or personal connections or networks that are created by repetitive and long-term transactions are used. There can be many other forms of signals and arrangements. All of these in their totality are defined as the "institutional arrangements." In other words, all of the non-price or quasi-price mechanisms which enable transactions to take place in the incomplete market are the "institutional arrangements."

Institutional arrangements in Asian villages are a dynamic process and change over time. An institutional arrangement may not always function well. However, the emergence of proper institutional arrangements to replace the old ones can be due to technological innovation, the introduction of new products, a change in relative price, or the reform of the legal and legislative framework. Our concern is to analyze the institutional arrangements within the dynamism of agricultural development in a region which has succeeded in carrying out rapid economic growth. As mentioned earlier, the economics of information and the new institutional economics have provided the theoretical framework for analyzing economic transactions under market incompleteness. Within this framework the existence of transaction costs, moral hazards, and adverse selections are seen as the three important defects that occur under market incompleteness. However, the obvious symptoms of these defects, particularly moral hazards and adverse selections, have hardly been observed in the studies presented in this special issue. This observation is consistent with our view that the institutional arrangements deter the emergence of these defects. Institutions can have negative or conservative effects, nevertheless proper institutional arrangements can be produced which can improve the economic welfare of rural people. The interpretation of "institutional arrangements" can vary from an institution in the ordinary sense to the market itself, or the legal framework which supports the market mechanism.

The three main elements forming the market are: the rules of economic transactions, the individual players or organizations who conduct transactions, and the physical infrastructure of the market. We pay attention in this study to the first two elements as institutions.

Implications

The major findings of the studies in this special issue can be summarized in the following four points.

- Market incompleteness does exist and institutional arrangements to cope with this problem can be observed in Asian villages. This incompleteness, particularly of the factor markets, is related to the social structure of agrarian societies. Therefore changes in the institutional arrangements are very likely to accompany a transformation in the social structure on which the institutional arrangements are based.
- The sphere of a certain institutional ar-rangement is restricted by its social background which limits its effect range to a village, a specific group, an organization, or a social relationship.
- An existing institutional arrangement is rational under the given existing constraints. Our case studies did not observe any marked inefficiency in resource allocation given the various constraints that an institutional arrangement had to cope with. This means that autogenous and/or traditional institutional arrangements are not simply outmoded affairs harmful to the process of development, and they should not be rejected indiscriminately. Rather they could be capable of contributing to agricultural development and achieving a well-arranged market economy.
- The forms of institutional arrangements are changing in rural Asia under the process of agricultural development. This change is widening the sphere of more uniform institutional arrangements, and price signals are becoming much more important than they

were in the past. Rural Asia is responding to this change and is adapting effectively to it.

These findings have important implications for structural adjustment policy that takes a "market-friendly approach." A fundamental idea of structural adjustment policy is that the presence of a complete market is a prerequisite. However, such a complete market hardly exists in Asian agriculture. Nevertheless, policies based on this idea have been implemented and the diversification of agriculture has somehow taken place. This must be due to the fact that some traditional and/or autogenous mechanisms have complemented or supplemented the incompleteness of the market. We regard this kind of change and response as also a form of "institutional arrangements." The agrarian societies with capable institutional arrangements have succeeded in responding to diversification, and the institutional arrangements must have played a substantially important role in supporting agricultural development.

The institutional arrangements may not always work out in a favorable way. There is always a possibility that the arrangements will not take form, or that an old institution will suppress change. Where a desirable institutional arrangement is lacking, structural adjustment policy cannot be carried out effectively. To conclude, if the recent market-friendly approach takes into consideration the capability of the institutional arrangements in each country and within the local areas of each country, it would substantially increase the applicability of structural adjustment policy.

Job Opportunities in Agricultural Industry

The transfer of labor from agriculture to industry and

service employment that typically accompanies economic development was stifled in China for many years.

The household registration system instituted in the 1950s made clear distinctions between rural and urban residents. Rural residents could not freely move to urban areas, so there was no outlet for growth in the rural labor force. When China began its economic reforms in the late 1970s, nearly all rural residents were engaged in subsistence farming because there were few opportunities for rural nonfarm employment.

Rural nonfarm employment has grown rapidly over the past two decades, but nearly half of China's labor force is still engaged in agriculture. Farm earnings are low because each worker is limited in both land and capital. In 1999, rural per capita income in China was just $360, reflecting low earnings in agriculture, which accounted for 60 percent of rural income.

China's high agricultural share of employment and low ratio of land per farm worker are not unusual among developing countries in Asia. However, middle-income and developed countries in Asia have 10 percent or less of their workers in agriculture and ratios of land per worker are much higher than that of China (table 1). To raise per capita output and income in farming, China's economy needs to shift large numbers of farm laborers to employment in industry and services, where earnings are higher. Creating nonfarm opportunities will become an even higher priority as China's agriculture sector becomes more integrated with the world economy.

Table 1.

Agricultural labor force share and land per worker, selected countries and regions, 1999

Country/region	*Agricultural share of labor force Percent*	*Land per agricultural worker Hectares*
Laos	76	.44
Cambodia	70	.82
Vietnam	67	.21
Africa - developing	60	.83
India	60	.61
Thailand	56	.70
Bangladesh	56	.21
Indonesia	48	.36
Pakistan	47	.87
China	47	.39
Sri Lanka	46	.23
Philippines	40	.45
Mexico	21	2.84
Latin America and Caribbean	20	3.02
Malaysia	19	1.03
South Korea	10	.71
Europe	9	9.36
Taiwan	2 8	1.10
Australia	5	107.34
Japan	4	1.63
Canada	2	116.82
United States	2	58.46

China's policymakers and planners need better information about rural employment. Until recent years, "rural" and "agricultural" were essentially synonymous, and the number of farmers could be easily counted as the number of persons with agricultural household registrations. However, during the 1990s, many agricultural people found jobs in cities (though still mostly illegal) and in growing rural nonfarm enterprises. The number of people with agricultural registrations (over 900 million, including children and retired people) now overstates the farm population.

Estimates of the number of rural people employed in farming and rural nonfarm industries vary widely. In 2000, official statistics reported a rural agricultural labor force (including forestry and fishing) of 328 million and a rural nonagricultural labor force of about 170 million (table 2).

Table 2. China employment and growth by sector, 1990-2000

Sector	*Employment 2000*	*Growth 1990-2000*
	Million	
Agriculture, forestry, and fishing	328	-5
Rural township and village enterprises	128	35
Rural private enterprises	11	10
Rural self-employment	29	14
Urban employment	213	47
Total	710	101

Some argued that official statistics published in China's statistical yearbooks underestimate rural employment in nonfarm industries and overestimate farm employment by perhaps 100 million. However, China's 1997 agricultural census, the first attempt to obtain a comprehensive nationwide count of rural employment, reported an even larger number of 425 million persons primarily employed in agriculture, forestry, and fishing and 136 million rural persons employed in nonagriculture (including 57 million working primarily in urban areas).

Where Are the Jobs?

China faces an enormous task in moving labor out of agriculture. Johnson estimated that China would need to create 15 million jobs per year over three decades to reduce its farm employment to 10 percent of the labor force (about the level of South Korea and Taiwan). This rate is nearly three times the average 5.9 million workers per year that transferred from agricultural to nonagricultural activities from 1978 to 2000, according to China's National Bureau of Statistics. A 20-percent agricultural labor force share (similar to that of Malaysia) is probably more realistic but would also require China to accelerate job growth over current rates. Rural job growth faces a number of challenges.

Rural township and village enterprises (TVE) have absorbed much farm labor but are now trying to raise productivity and workforce quality, which tends to reduce hiring. TVE employment grew by 35 million during the 1990s but fell by 7 million from 1996 to 2000. Private enterprise and self-employment in rural areas added 24 million jobs during the 1990s but still account for less than 10 percent of rural employment.

From 1990 to 2000, urban areas provided 46 percent of China's employment growth-much higher than their 30-percent share of the country's population in 1990. According to China's 1997 agricultural census, 40 percent of rural residents working in nonagricultural activities worked in urban areas. Thus, much of the labor moving out of agriculture will likely find jobs in urban areas. As economies develop, most employment growth is in service sectors, and these jobs tend to cluster in urban areas, especially large cities, which are geographic centers for trade and incubators for high-tech industry. Labor-intensive low-skill services will be a logical sector for absorption of China's rural labor force. Manufacturing industries in both rural and urban areas will be under pressure to keep labor costs competitive and increase capital, skill, and productivity per worker. Textile manufacturers, an important employer of rural labor, have already undergone significant consolidation and downsizing.

Institutional Reforms May Aid Mobility

Studies of rural migration in China find that migration tends to follow networks established by village members. Migration is largely temporary, circular, and over short distances. Migrants tend to be young unmarried adults who are members of households with limited farmland. Remittances sent home by migrants are an important source of capital for rural households. While mobility of rural labor has increased at a rapid pace, it is still constrained by limits on urban migration, grain procurement obligations and land-tenure systems, and lack of rural credit markets. As China relaxes some of these barriers to migration, what we know about migration from previous studies may become outdated.

Increased education will make China's rural workers more productive and employable. Rural education levels have increased dramatically since the 1950s, especially for women, but many observers believe rural schooling levels are now in decline. Rural local governments must fund their own schools, and many poorer areas have little resources to support education. Rural residents view schooling as a means to migrate to an urban job. In other Asian countries, rural residents see better educational opportunities in urban areas as an important motive for moving to cities. Studies of the impact of education on farmer productivity have found mixed results. Persons with the highest level of education and skill are the most likely to enter nonagricultural work, leaving the less skilled in farming. However, a general rise in rural education seems likely to improve the ability of farmers to understand and adopt new technologies and process market information. Part-time farming in China is becoming much more prevalent as farm household members commute to offfarm jobs. New institutional innovations in land tenure may allow farm households to maintain "ownership" of their land while devoting their labor to nonfarm work. In Taiwan, for example, over 80 percent of farm households have at least one member working offfarm. The fairly uniform distribution of small land plots among farm households in China may become more skewed as labor migration and consolidation of land occurs in some villages. Marketing arrangements and contract production for processors that need steady supplies of uniform products are becoming more common, which may speed up the commercialization and modernization of farming in China, turning "peasants" into modern farmers.

Food Insecurity

In addition to measurements of children's height and weight, other indicators of health and sanitary conditions can help identify and respond to nutritional problems. The nutritional indicators used are percentage of children under five years old who were wasted, stunted or underweight at the time of the most recent survey. The health indicators are life expectancy at birth, the mortality rate for children under five and the percentage of the population with access to adequate sanitation. The occurrence of high levels of all these indicators together obviously indicates severe health and nutritional problems. Conversely, when all indicators are low, the situation is much more benign. And, indeed, the performance of individual countries bears this out. Among the approximately 80 countries that have been examined, nine achieved excellent scores for all indicators. In all nine countries, the prevalence of undernourishment, as estimated based on food availability and access, was also moderately or very low. At the other end of the scale, 16 countries scored very badly on all six indicators. All 16 also suffered from moderately high or very high prevalence of undernourishment.

In general, the impact of food intake, health conditions and care practices on children's nutritional status can be understood best by analysing levels and changes in relevant indicators for individual countries over time.

Children's nutritional status tends to be better in countries where people get more to eat, live longer and have better sanitary facilities. The fact that high percentages of children are underweight even in some countries where prevalence of undernourishment is low or access to sanitary facilities is very high confirms that nutritional status is determined by a combination of many different factors. Improved understanding of the interplay among these factors requires more systematic data collection and more sophisticated analysis.

Most hungry and malnourished people around the world are likely to be found in vulnerable areas -locations where environmental, economic and other factors expose them to a high risk of impoverishment and food insecurity. But the specific factors that determine vulnerability vary from one location to another. In order to understand these differences and adopt suitable responses, vulnerable areas must be analysed at national and subnational levels.

Poverty and Malnutrition

Poverty and malnutrition remain endemic in many coutries of the world. For example in Bangladesh, an overwhelmingly agrarian country where most rural households do not own land and few other opportunities to earn an income exist. At barely 2 000 calories per person per day, food availability falls short of meeting basic requirements. Some areas of the country still face the risk of famine. Overall, the rate of undernourishment is very high (37 percent), as is the prevalence of underweight, stunting and wasting among children. Rates are high throughout the rural areas that are home to 80 percent of the Bangladeshi population. More than 60 percent of rural households are functionally landless and there is very little income diversification. The level of vulnerability is aggravated by floods during the annual monsoon season, cyclones and drought.

Another example is Venezuela . With substantial income from oil production it is a highly urbanized country with enough food available to meet requirements. But poverty remains widespread, affecting almost half the population. And up to one-third of the children in some impoverished rural states show signs of repeated or prolonged periods of undernutrition. On a national scale, a moderately low 15 percent of Venezuela's population are undernourished and rates of stunting, wasting and underweight among children are all low. As the map indicates, however, between 20 and 36 percent of children suffer from stunting in several states in the south and east of the country where around two-thirds of the population fall below the poverty line.

Rising unemployment and high inflation have exacerbated the nutritional problems of the rural poor and especially indigenous people. More indepth analysis is needed to identify pockets of poverty and malnutrition in the cities where around 85 percent of Venezuela's population live.

Monitoring the number of undernourished people provides a way to gauge how much progress has been made and how much remains to be accomplished. It keeps attention focused on the target. But knowing the number is not enough to guide action. For that, more specific information is needed about the groups of people who are at high risk of food insecurity and the factors that create this risk.

Environmental Xonstraints

Many environmental factors - some natural, some the consequence of human activities - can limit agricultural potential and create vulnerability. Those shown on the map include extremely dry or cold climates, poor soil, erratic rainfall, steep slopes and severe land degradation. A full analysis would also need to incorporate other human factors such as pollution and landmines.

Interestingly, zones where rainfall is erratic typically have good agricultural potential. This is because the highest variability in rainfall occurs in high rainfall areas, which are often conducive to agriculture. In fact, worldwide, many of the most fertile soils and favourable climates are located in areas subject to environmental extremes, such as tropical cyclones. In such areas, storms and floods can wash away a season's harvest, along with the fertile topsoil, unless precautions are taken.

People have naturally tended to concentrate in the areas with high production potential, characterized by beneficial climate and soils. Often this concentration of people has led to overexploitation of natural resources, which, combined with the inherent rainfall variability of high-potential areas, has resulted in land degradation. Indeed, many areas where land is severely degraded today, such as the East African highlands, were once highly suitable for agricultural activity. So environmental risks exist even in areas shown on the map as having medium or high potential for agricultural production. In many parts of the world, most often industrialized countries, these risks are being successfully managed through investment and use of appropriate technologies. Where this is not the case, people's livelihoods may be adversely affected, leading to vulnerability and food insecurity.

Dynamics of Change

Analysis of progress in reducing hunger rarely reveals startling changes from one year to the next. Occasionally a major shock of nature or history - such as a cyclone, economic slump or war - may trigger a food security crisis. But the impact of such events tends to be transitory, reflected in a sharp spike in the numbers that does not alter the long-term trajectory. Reaching the World Food Summit target will require successful efforts to change the long-term trends and to understand the many factors that determine them.

Conditions vary considerably from one place to another and so do the combinations of factors that leave particular population groups vulnerable to poverty and hunger. Usually many different demographic, environmental, economic, social and political elements are involved. In the follow-up to the World Food Summit, considerable effort has gone into developing effective methods and indicators to monitor these factors. Some preliminary results of this effort are presented in this section by focusing on eight countries that have registered particularly significant changes in prevalence of undernourishment since 1980.

Economic Growth and Food Security in Asia

A period of rapid economic growth has resulted in major gains in food security across most of Asia and the Pacific. Cambodia, where the proportion of undernourished dropped from 62 to 33 percent between 1980 and 1996, led the way. Many other countries in the region also showed strong reductions, including China, India, Indonesia, Myanmar, Nepal, Pakistan and Viet Nam. Under nourishment increased in only two countries - Mongolia and the Democratic People's Republic of Korea. In the latter, it rose from 16 to 48 percent.

The financial crisis may have blunted progress in some countries but the overall trend remains positive. Mixed picture of progress and setback characterizes Latin America and the Caribbean Both the levels and the trends of undernourishment vary considerably in Latin America and the Caribbean. In most South American countries, levels are already low or rapidly declining. In Central America, on the other hand, levels are increasing in several countries, although Honduras did register the strongest gains in the region by bringing its prevalence of undernourishment down from 31 percent to 21 percent. In the Caribbean, Cuba's setback, with the share undernourished rising from 3 to 19 percent, was in many ways typical of several of its island neighbours, which have experienced increases in undernourishment since 1980.

Most countries in the Near East and North Africa have already achieved very low levels of undernourishment. Indeed the region accounts for 10 of the 14 developing countries where undernourishment affects less than 5 percent of the population. Morocco's decline, from 10 to 5 percent, represented the best progress among this group of good performers. Significant increases occurred only in Afghanistan, where the proportion of people who are undernourished shot up from 33 percent in 1980 to 62 percent in 1996, and in Iraq, where the share rose from 4 to 15 percent over the same period.

Several West African nations register gains but problems grow worse elsewhere in Africa With Ghana leading the way, eight countries in West Africa reduced hunger significantly between 1980 and 1996. Indeed the five countries worldwide that scored the biggest gains were all from this subregion. The picture was very different in Central, East and Southern Africa, however. There, the proportions and numbers of undernourished people generally increased. Burundi suffered the largest increase, with the proportion of undernourished people rising from 38 percent to 63 percent between 1980 and 1996. But 13 other countries in Central, East and Southern Africa also showed large increases.

Threats of floods, drought, war and financial collapses

The most recent global estimates for the number of hungry people in the world are based on data through the end of 1997. Since then, a series of environmental, economic and political shocks have disrupted agricultural production, driven millions of people from their homes and fields and swelled the ranks of the poor in places as farflung as Honduras, Indonesia and Kosovo.

Floods Shocks

In 1998, weather patterns associated with El Niño parched crops in some regions, drowned them in others, and hammered Central America with the devastating winds and rain of Hurricane Mitch. More than 9 000 people died and nearly 3 million were left homeless when Mitch cut across Central America between 26 October and 1 November. Floodwaters and mudslides swept away buildings and roads. Fields of maize, beans and sorghum were laid waste. Honduras and Nicaragua were hardest hit, with Honduras losing more than half its staple maize crop. Losses in coffee and other export crops added up to an estimated US$ 480 million. The region faced both an immediate humanitarian crisis and a long-term threat to food security.

The international response was swift and effective. The UN Office for the Coordination of Humanitarian Affairs (OCHA) sought US$ 150 million for relief and rehabilitation. The World Food Programme (WFP) responded immediately with food aid. FAO organized the distribution of seeds, fertilizers and hand tools to help farmers in the hardest-hit rural communities feed their families as soon as possible.

Work has now begun on rebuilding the agricultural sector. Improved farming practices are being introduced that will reduce environmental degradation and vulnerability to future disasters. Hurricane Mitch could not have been avoided. But manmade factors greatly increased its destructive impact. The most deadly mudslides swept down slopes denuded by deforestation and cultivation of marginal lands. Flooding was aggravated by poor watershed management.

The most affected countries are attempting to address structural problems that contributed to the disaster by examining land tenure practices, supporting reforestation projects and providing training in watershed management. International financial institutions have made commitments to provide up to US$ 5 300 million over the next four years. While more deeply rooted economic problems are likely to persist, the recovery programme should prevent increases in food insecurity and vulnerability throughout the region.

Even as hope takes root in Central America, however, severe flooding in several countries of East and Southeast Asia and a withering drought in the Near East have threatened to set back recent progress toward the World Food Summit goals. For the Near East region as a whole, production of cereals in 1999 is expected to fall 16 percent below the previous harvest. But the losses from the worst drought in decades will be far greater in Iran, Iraq, Jordan and Syria.

Several thousand farmers, smallscale herders and landless rural households who have lost their means of livelihood will require assistance to tide them over until next year. As in Central America, however, there is currently no evidence to suggest that these setbacks will be more than transitory in either Asia or the Near East.

War and Complex Emergebcies

From Angola in Southwestern Africa to Kosovo in Southeastern Europe, many parts of the world ravaged by war and civil strife face complex emergencies requiring many types of humanitarian assistance. (The map above shows these countries, plus others requiring exceptional food assistance.) The long-simmering conflict that boiled over in Kosovo in the spring of 1999 attracted the most international attention and involvement. This conflict created enormous human suffering and laid waste much of the province's productive infrastructure. But the outpouring of humanitarian aid has been more than sufficient to stave off hunger for at least a year.

Whether levels of undernourishment will increase after the aid is withdrawn will depend on the success of rehabilitation and recovery programmes currently being designed and put into operation. The situation in Angola is much more troubling. There, renewed fighting has driven nearly 2 million people from their homes. Many have fled from the countryside into besieged cities and towns where they must depend on airlifts of food aid for survival. As of mid-year, food distributions were able to meet only 60 percent of the needs and FAO experts warned of a catastrophe, with the threat of mass starvation among displaced people.

Europe, 14 countries and one subregion faced complex humanitarian emergencies in 1999. The total requested under OCHA appeals for such emergencies increased by 20 percent over 1998. But the total amount of humanitarian aid pledged as of July 1999 has declined slightly and a far higher percentage is going to Southeastern Europe. Inevitably, that means less is available to meet essential human needs elsewhere. Afghanistan, Angola, Guinea-Bissau, Sierra Leone, Somalia, Sudan and Uganda are all countries where substantially lower percentages of total needs have been covered this year than in 1998. In these countries, the spread of undernourishment is inevitable.

Financial Crisis

When several Asian banking systems collapsed in 1997 there were fears that poverty and food insecurity would soar in Indonesia, Malaysia, the Philippines, the Republic of Korea and Thailand and that their "tiger economies" would take years to recover. These countries had achieved impressive economic success and succeeded in reducing poverty substantially over two decades. When financial systems collapsed throughout the region, many of these gains were reversed. Businesses closed and large numbers of office and shop workers lost their jobs.

The shock waves proved less destructive than first feared, however. By the beginning of 1998, after dramatic falls in their exchange rates and purchasing power, most of the affected countries started to show signs of at least partial recovery. Although poverty and hunger undoubtedly increased somewhat throughout the region, for most countries the crisis was short-lived. Indonesia has been the exception. In just one year, the fourth most populous country in the world saw its currency fall to one-fifth its previous value. The inflation rate soared to 50 percent and food prices rocketed. At the same time, the worst drought in 50 years was followed by excessive rains during the dry season. Bad weather and reduced use of imported fertilizers and pesticides caused a slump in local rice production. Imports increased, but not enough to make up the shortfall.

Job losses and inflation hit urban residents particularly hard. A household survey conducted by the World Bank and the Ford Foundation found that urban households spent 34 percent less per person in 1998 than they had in 1997. In rural households, expenditures declined by only 13 percent. In many cases, families that had been poor before the crisis actually benefited from increased business in the informal sector. But the newly unemployed had to sell off assets and deplete their savings in order to cope. As the crisis deepened, many hard-hit families turned to relatives in the countryside for help, particularly in Java. This increased the pressure on those who still had jobs and assets, but now had to feed a growing number of dependants.

The impact of the economic collapse on nutritional status showed up quickly. Most households continued to consume normal quantities of rice, but cut back on more expensive foods, such as meat, fish and eggs. Reduced consumption of foods rich in protein and micronutrients led to sharp increases in the numbers of wasted children and anaemic mothers. Current estimates suggest that the proportion of Indonesians who are undernourished almost certainly doubled from 6 percent of the population in 1995/97 to 12 percent in 1999. And the current figure could be as high as 18 percent. This means that the economic crisis has added 10 to 20 million people to the ranks of the undernourished in Indonesia alone. Recovery has begun. But it will take time for Indonesia to create the conditions needed to bring undernourishment back to the levels of before the crisis and continue on the path of progress towards the Summit target. The task has been complicated by events in East Timor, where tens of thousands of people were driven from their homes following the referendum on independence. Humanitarian assistance was mobilized, but if the conflict exacerbates political and economic instability in Indonesia, there is a grave risk that the food security situation could get worse before it gets better.

Commercialisation of Small Farm Agriculture

The crux of agricultural development in India is to maintain a rising level of foodgrain production and a reduction in rural poverty by speedy growth of income and employment. The new seed-fertiliser based technology has helped the farmers achieve the first goal of food security, because priority for foodgrain production is rightly compatible with subsistence orientation of the majority of the farmers - especially marginal and small farmers. In the past, it was believed and in fact verified by several studies that the new technology is scale neutral and the poor would share at least equally with rich in the gains from the technology. This has not happened and poverty persists. It is a paradox that a record foodgrain production of 186 million metric tonnes in 1994-95 coexists with nearly a third of the country's population living below the poverty line. Indeed, the economic inequalities have widened over the years, revealing that integrating capitalist agriculture with welfare goals is more formidable than earlier thought of. One reason is the lack of adequate attention in policies and programmes for agricultural development to the need for rapid commercialization of agriculture. The history of the presently developed economies shows that modernisation, monetisation and integration are the essential ways of agricultural development. The new technology helped modernization. Further, the increasing demand for money - to purchase most of the input offfarm, leads to increasing sensitivity of farmers to relative prices of agricultural commodities and that helped monetization and commercialization. The next step is to ensure the transfer of resources (generated surpluses) -wage goods, raw materials and investible funds - from agricultural development and improve the capacity of fanners to adopt new agricultural technology.

Therefore, it would increase the income of the farmers and contribute to creating employment for the landless. This is the process and theory of 'Integrated Rural Development'. The farm size, direction and pace of this transfer are important. Indian experience is that the process has bypassed a vast majority of the producers in agricultural sector and the progress in integration is highly inadequate. While modernisation and monetization have helped agricultural growth - largely in foodgrain production and commercialization on a limited scale, the process has failed to encompass the marginal and small farmers.

Marginal and Small Farms

By their numerical strength, the marginal and small farmers form the majority of cultivators in the country as a whole and expecially in Tamil Nadu and Madurai district. The pattern of size distribution of farms is similar - the predominance of marginal and small farms in both number and area operated by them. Compared to all India, their predominance is much greater for Tamil Nadu and Madurai district (Table 3).

Therefore, a sample of 200 farms was selected in Madurai district by a three stage random sampling method - the stages being CD blocks, villages and farms and by distributing the sample among the villages in probability proportion to the number of farms in each village to the total number of farms in all the 20 sample villages in five C.D. blocks. These sample farms were post-stratified into three size groups based on this operational area viz., small (< = 2 ha); medium (2-4 ha) and large (>4 ha). By definition therefore, small group included marginal farms and the large group included medium farms and the group of semi-medium in the secondary data corresponded to the medium size group of the sample.

The distribution of sample farms among the size groups and area operated by them are shown in Table 4.

Table 3.

Number and Operational Area of Farms - Size group wise

	All India		*Tamil Nadu*	*Madurai*	*Dt Size Group*	
	No. million	*Area m.ha*	*No. '000*	*Area' 000 ha*	*No. ' 00*	*Area ' 00ha*
Small	74.07	47.75	47.75	3790	2816	2051
(<=2ha)	(76.23)	(29.02)	(87.68)	(48.61)	(88.96)	(56.96)
Semi medium	13.25	36.67	649	1778	231	708
(2-4 ha)	(13.64)	(22.28)	(22.28)	(22.81)	(7.30)	(19.66)
Medium	7.92	47.14	261	1508	101	580
(4-10 ha)	(8.15)	(28.84)	(3.39)	(19.34)	(3.19)	(16.10)
Large	1.92	33.00	39	720	17	262
(>10ha)	(1.98)	(20.06)	(0.51)	(9.24)	(0.54)	(7.28)
Total	97.16	164.56	7707	7796	3165	3601
Per farm	_	1.69	_	1.81	_	1.14

Table 4

Distribution of Sample Farms-size group wise

Size Group	*Total*		*Total Area*		*Per Farm Area (ha)*		
	No.	%	*Ha.*	%	*Mean*	*Min*	*Max*
Small (<=2 Ha)	168	84.00	124.49	53.94	0.741	0.04	2.00
Medium (2-4 Ha)	20	10.00	45.40	19.67	2.270	2.20	3.60
Large (>4ha)	12	6.00	60.91	26.39	5.076	4.06	14.00
Total	200	100.00	230.80	100.00	1.15	0.40	14.00

The striking similarity between the primary and secondary data for Madurai district in the percentage share of operational area of the three size groups in the total area of all the farms cannot be missed. Consequently the average size of operational area of a farm is 1.15 ha in the sample and 1.14 ha in the data for the district. Thus the sample is adequately representative of the district and allows generalisation of the results. In all the size groups, average area of the farm was closer to the lower limit of the range rather than the upper limit.

Cropping Pattern

The concept of commercialization refers to the increasing share of the marketable surplus in the total farm business income of the farms. This share can be increased by (i) generation of more marketable surplus in subsistence oriented foodgrain crops; (ii) by increasing production of market oriented cash crops and other products (collectively known as high value adding enterprises) and (iii) both. An increase in foodgrain production through improvement in productivity of land and diverting area for the high value adding (HVA) crops or other enterprises will ensure food security and commercialization simultaneously.

Therefore the cropping pattern of the farms will show the share of high value adding crops. Groundnut, onion, chillies, sugarcane and cotton are the value adding (cash) crops cultivated by the farmers of the district, while paddy, jowar and blackgram are the food crops raised by them. The share of the non-food crops in the gross cropped area of the farms is shown in Table 5.

In the total gross cropped area of 297.05 ha , cash crops accounted for 64.03 ha (21.56 percent) only showing the farmer's preference to food crops - a sign of subsistence orientation. The share of the food crops in gross cropped area was as high as 90 percent in small farms as compared to 59.75 percent in large farms.

Therefore commercialization through production of products for market (cash crops) was very small i.e., less than 10 percent in total area under cash crops. The large farms were just 12 (six percent) in the sample of 200 farms, but their share in area under the cash crops was 53.60 percent. Therefore the commercialization of agriculture would need attention to small farms, to encourage area under cash crops.

Productivity of crops

The average productivity (in kg/ha) of the crops raised in the sample farms is shown in Table 6 for a comparative study. In jowar, bajra, blackgram and groundnut (rainfed crop), the small farms had the highest productivity. In other crops excepting sugarcane, the productivity of small farms was not significantly different from that in other groups. Only in sugarcane average productivity per hectare was just 101 tonnes/ha as compared to 140 tonnes/ha on large farms.

Therefore, productivity was not a serious constraint for commercialisation of small..farms if the marketed surplus were not different for the size groups. By the time of enquiry (April '95) the sample farmers had completed the sale of crops raised in the previous years. Therefore their marketed surplus was estimated and expressed as percentage to total production.

Table 5.

Share of Cash Crops in Gross Cropped Area

Size Groups	No.of Farms	Gross Cropped Area (ha)	Cropping intensity %	Area under (ha) Food Crops	Cash Crops	*
Small	168 (100.00)	152.779	122.73	137.68 (90.11)	15.11 (9.89)	23.61
Medium	20 (100.00	509.00	129.95	44.40 (75.25)	14.60 (24.75)	22.80
Large	12	85.26 (100.00)	139.97	50.94 (59.75)	34.32 (40.25)	53.60
Whole Sample	200	297.05	128.70	233.02	64.03	100

Table 6. Productivity of Crops (1993-94)

Crops	Productivity in Kg/ha, in		
	Small farms	Medium farms	Large farms
1. Paddy	3089	2995	3286
2. Jowar	1336	1217	1180
3. Bajra	1205	1110	1070
4. Black gram	786	552	654
5. Groundnut(R)	1332	1280	1295
6. Cotton (R)	249	279	281
7. Onion	820	864	848
8. Chillies	790	827	830
9. Sugarcane (t/ha)	101	115	140

Marketed Surplus

Estimates of average marketed surplus as percentage to production are presented in Table 7.

Table 7. Marketed Surplus in Crops Produced

Crops Raised in 1993-94	Marketed Surplus (percentage) on		
	Small Farms	Medium Farms	Large Farms
1 Paddy	54.15	68.00	76.13
2 Jowar	50.20	57.50	96.70
3 Bajra	63.45	89.38	99.85
4 Blackgram	48.74	46.53	47.77
5 Groundnut (R)	100.00	100.00	98.80
6 Cotton (R)	100.00	100.00	100.00
7 Onion	98.70	99.20	100.00
8 Chillies	100.00	100.00	100.00
9 Sugarcane	100.00	100.00	100.00

It is readily seen that small farmers had the lowest percentage of marketed surplus in food crops, with only exception of blackgram, but they had 90 percent of GCA under them. The area under cash crops was just 10 percent but their marketed surplus was 100 percent with one exception of onion where it was 98.70 percent. The inference is that production in small farms is more subsistence oriented than market oriented. This is a constraint for their commercialization.

The farmers maintained one or two cows/she-buffaloes, work bullocks, sheep and poultry birds. Income from livestock supplemented the income from crops. But the size of livestock depended upon the availability of family labour to attend them. As members of the family of small farms worked as wage earners off-farm and non-farm, the livestock maintenance was on a small scale and brought very little income to the farms. It was not a commercial activity. This can be seen in the source-wise income of the farms.

Farm income

The income of the farm included income from crops (sale proceeds plus imputed value of produce retained by the family), livestock, off-farm income from wages earned by the members of the family going for work in other farms and income from other sources such as non-farm employment and property income. In all the size groups, income from crops exceeded sixty percent, but it was least on small farms. Income from livestock was negligible (0.92 percent) on small farms, about 15 percent on large farms and 21.42 percent on medium farms.

The small farms would be able to maintain livestock with the help of straw from grain crops that dominated the crop pattern, but they had not, largely because members of the family went out to work on other farms as shown by the large (35.74 percent) wage income from such works. Large farms had very little family labour and therefore, their attention was mostly on crops and cows and poultry were maintained mostly to meet the home needs. It was the farms in the medium size group that had family labour which preferred maintaining livestock rather than going to work off-farm. Hence, income from livestock had a share of 21.42 percent of gross income, while off-farm wage income was just 1.40 percent.

Other sources of income included non-farm employment of family members, properties and other miscellaneous sources. The share of this source increased that too very significantly as the farm size increased. Small farms were seen to be better than the medium farms in their operational efficiency as shown by the gross income/cost ratio and net income. Then operational ability of the farmers is not a constraint for their commercialization.

Commercialization

In most of the recent literature, commercialization is approximated by the relative share of cash crops in the GCA. A more accurate measure of commercialization of farming (as opposed to commercialization of crop production)-especially in the context of a diversified agriculture including livestock and allied activities-is the percentage share of income from the sale of farm products in the total value of production. What did not enter the market would be retained for use on-farm and its value had to be imputed. The estimates of value realised from sales and imputed values of the retained quantities are presented in Table 8.

Table 8. Degree of Commercialization Compared for the groups

Sources of Income		*Small Farms*	*Medium Farms*	*Large Farms*
Gross income from	Crops	104.12	29464.36	78360.44
	Livestock	159.38	9725.30	18311.16
	Total	10647.50	39189.66	96671.60
Sale Proceeds from	Crops	3081.41	18768.80	66144.05
	Live-stock	53.15	5115.51	12836.01
	Total	3134.56 (29.44)	23884.31 (60.95)	78980.06 (81.70)
Imputed Value of Retention in	Crops	7406.71	10695.56	12216.39
Livestock		106.23	4609.79	5475.15
Total		7512.94 (70.56)	15305.35 (39.05)	17691.54 (18.30)

The percentage value of farm products sold out to the total value of production was just 29.44 percent on small farms, while the value of retained products accounted for 70.56 percent, thus clearly revealing that small farms were subsistence oriented. In contrast, the share of sales in total value product was 61 percent on medium farms and 82 percent on large farms showing them to be largely market oriented. What would explain this situation? Opinion of the farmers was analysed and it revealed three major causes for their reservation on expanding area under cash crops and lack of interest in cultivation of high value adding horticultural crops such as vegetable, fruit and flower crops. The opinion of the small farmers is presented in Table 9.

Table 9. Reasons behind allotting restricted area under high value crops-opinions of small fanners

Reasons	*Seasonal Cash Crops*		*Horticulture Crops*		*Allied activities*	
	No.	*%*	*No.*	*%*	*No.*	*%*
Non-availability of credit	97	57.50	21	12.50	87	51.80
No Buyers			86	51.20	56	33.30
Highly Risky	25	14.90	83	49.40	-	-
Huge Investment Not Possible	-	-	120	71.40	-	-
Price is not satisfactory	27	16.10	20	11 .90	-	-

In seasonal cash crops which included groundnut, cotton, sugarcane and chillies, the nonavailability of credit was the most important constraint, followed by risk and unfavourable price. In livestock products of dairy, poultry and piggery, non-availability of creditand want of buyers (i.e., non availability of market outlet within their easy reach) were the reasons for limiting the size of those enterprises. In horticultural crops which included vegetables, fruits and onion, the need for making huge investments with production lag and the risk involved due to their bulkiness and perishability and also absence of market were the reasons reported by the farmers, as the constraints.

The inability of the farmers to invest, reflected on the internal credit rationing, while their complaint of non-availability of credit would show the external credit rationing. Therefore, credit rationing and unfavourable market emerged to be the two major limiting factors for the commercialization of small farms. Interaction with officials of the commercial banks and Primary Agricultural credit Banks showed that they considered small farms as unviable units even for short term lending, not to speak of long term development credit. This statement was verified by analysing the ability of the farmers to repay the credit by comparing their income and expenditure levels and possible savings per annum.

Problem focus

On an average the family of a small farm spent Rs. 10,380 per year, while it was Rs. 21,242 for medium farms and Rs 65,516 for the large farms. At this level of expenditure, there was little scope to reduce the consumption expenditure of a small farm family, which is neatly one half of the medium size group and one sixth of large farms. When this level of expenditure was compared with average annual income of the farms, small farm had1 only a negative saving, while medium and large farms saved nearly 10 percent and 19 percent of their income respectively. So, bankers are right that small farms were not viable for credit. Only way for making them viable was to increase their income. It should be further emphasised that any increase in income would be spent to increase consumption first rather than saving, because the marginal propensity to consume is 0.82 (estimated from the sample data).

Therefore income would have to raise much more than what was required to remove the presently seen negative savings. To allow for atleast five percent savings, the average income of small farms should raise to Rs. 11,500. This was feasible with change in enterprise mix in favour of cash crops and improving productivity of the crops - both additional investment for which the average small farmer was not credit worthy. The problem is made more serious by two more facts. First majority among small farmers had farms smaller in size than the mean size, and thus they are problem farmers.

Secondly, there was wide variation in net income from crops and the coefficient of variation of the net farm income was 18.10 percent. This might bring more farmers within the risk net and credit would have a low level of probability of recovery. Therefore a low income - low investment cycle exists and it sustains rural poverty. How to break this cycle and help a cumulative growth path to small farmers? Some policy options are discussed below.

Table 10.

Income, Expenditure and Savings of Farm Households

Particulars	*Small*	*Medium*	*Large*
Net Income of Farms	10274.43	23736.87	812275.07
Consumption Expenditure	10381.48	21241.77	65516.44
Savings-Total % to Net	-107.05	2495.10	15758.63
income	-1.04	10.51	19.39
Discounted Savings(18%)	N.A.	3685.98	14722.08

Policy options

One way to break the cycle is to make credit available on the basis of prospective income or project based lending. The commercialization of farming significantly contributes to the net farm income. Size of the farm was not a constraint for it; only capital rationing and market size were the real constraint. The crop pattern seen in medium farms are technically feasible and economically viable.

Therefore alternate crop patterns may be identified and evaluated for their economic viability with the help of either partial budgets or linear programming technique to maximize aggregate net income of the farms. In the exercise, horticultural crops and allied activities may be the decision variables to be included. With the estimated costs and returns, plans with required level of savings may be identified and commended to the bankers. The advantage of this method is that the identified plans will win the confidence of both farmers and the lenders. Its major limitation is the cost and time involved in the exercise.

A second option is to encourage collective action such as group farming or cooperative farming, wherein farmers pool in their resources voluntarily to achieve larger scale of operation. Even medium size group farms were the net positive savers. Therefore, such a collective action is feasible. This is the approach suggested by Parthasarathy when he argues for alleviating poverty by a process of institutional reforms. Past experiences in institutional reforms such as cooperatives, regulated markets and C.D. blocks place a heavy discount on this strategy.

However, institutional reforms in the past were attempts to impose institutions on the farmers who never developed a sense of partnership in them. This weakness may be overcome by educating and guiding farmers to organise themselves rather than inviting them to join a sponsored institution. With this new approach, it is likely that cooperatives may emerge to be a practical way, of course with much of the present bureaucratic control discarded. It is of course a long term prospect.

A third option is the marketisation of the farm sector, with continued subsidy on inputs to the marginal and small farmers. This is consistent with the New Economic Policy which places faith on liberalization and invisible hand. However, the basic condition for such a policy to succeed is the presence of perfectly or near perfectly competitive conditions in the market. Such a condition does not exist and it will take a very long time to integrate farmers into the system. The required process of adjustment will ruthlessly eliminate poor and less organised and institutional cannibalism may be inevitable and that makes the process both painful and anti-thesis of liberalization for social uplift of the weaker sections- one being the small farmer.

Therefore only a combination of the above options will succeed. It must be recognised that as a social group, farmers are averse to collective action, unless the provocation is very strong. They also resist any imposition of institutions on them. However, if they have to survive in a liberalised economy, there is no alternative for an effective organisation for collective action.

Only recently there is some sign of lobbying among farmers. Exploiting this trend, they should be helped to evolve their own organisation. This will be the real remedy to the small farmers' problems. Yet, they have to be helped with project based lending for joint borrowing by a group of small farmers.

Bibliography

Alders, Carine and others (eds). *Linking with farmers : networking for low-external-input and sustainable agriculture*. London: Intermediate Technology Publications, 1993.

Americans for Safe Food, Center for Science in the Public Interest. "Taxing pesticides to fund research for sustainable agriculture : the Iowa model". Washington, D.C.: *Americans for Safe Food*, 1990.

Barbier, Edward B. (ed). *Economics and ecology: new frontiers and sustainable development*. London: Chapman & Hall, 1993.

Baudry, Jacques. "Interactions between agricultural and ecological systems at the landscape level". *Agriculture Ecosystems and Environment* 27 (1989): 119-130. Amsterdam: Elsevier Science, 1989.

Bentley, C.F. *World agriculture in the 1990's*. Ames, IA: Iowa State University, 1990.

Edwards, Clive A. "The importance of integration in sustainable agricultural systems." *Agriculture Ecosystems and Environment* 27, 1989.

Food and Agriculture Organization of the United Nations. Strategies for sustainable agriculture and rural development : new directions for agriculture, forestry and fisheries. Rome: FAO, 1994.

Gilson, J.C. *World agricultural changes: implications for Canada*. Policy study no.7. Ottawa: Renouf, 1989.

Kaiser, Harry M. and Thomas E. Drennen. *Agricultural dimensions of global climate change*. Delray Beach, FL: St. Lucie Press. 1993.

Parris, Kevin. *Agriculture and the environment : developing a set of indicators for use in agricultural policy analysis*. Paris: OECD, 1994.

Pierce, John T. "Agriculture, sustainability and the imperatives of policy reform". *Geoforum*, 24, 1993.

Singh, Naresh C. and Jacqueline Romanow. *Adaptive strategies and sustainable livelihoods in agricultural communities.* Winnipeg: International Institute for Sustainable Development. 1995.

United States Department of Agriculture Economic Research Service. *International agriculture and trade reports : western hemisphere.* Situation and outlook series. Washington, D.C.: Dept. of Agriculture, 1993.

Duvick, Donald N. "Responsible agricultural technology: Private industry's part." *Pro Rege* 16, 1988.

Freudenberger, C Dean. "The agricultural agenda for the twenty-first century." *Pro Rege* 16, 1988.

Walsh, John. "Preserving the options: food productivity and sustainability". Issues in agriculture no.2. Washington, DC: Consultative Group on International Agricultural Research, 1991.

Williams, Jeffery R. "Social traps and incentives : implications for low-input, sustainable agriculture. "*Journal of Soil and Water Conservation* 45, January/February 1990.

Wimberley, Ronald C. "Policy perspectives on social, agricultural, and rural sustainability." *Rural sociology* 58, 1993.

World Commission on Environment and Development. *Food 2000 : global policies for sustainable agriculture*. Geneva: WCED, 1986.

Index